Flame and Combustion

Flame and Combustion

J.A. BARNARD
DSc, PhD, BSc, DIC, ARCS, CEng, FIChemE
Sometime Newton Drew Professor of Chemical Engineering and Fuel Technology, University of Sheffield

and

The late J.N. BRADLEY
DSc, PhD, FRSC
Professor of Chemistry, University of Essex

SECOND EDITION

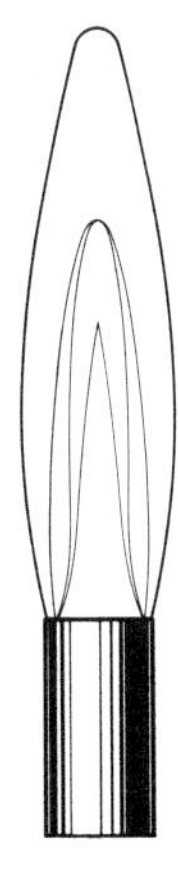

LONDON
NEW YORK
Chapman and Hall

First published 1969
as Flame and Combustion Phenomena *by*
Chapman and Hall Ltd
11 *New Fetter Lane, London EC4P 4EE*
Second edition 1985
Published in the USA by
Chapman and Hall
733 *Third Avenue, New York NY* 10017

Printed in Great Britain by J.W. Arrowsmith Ltd, Bristol

ISBN 0 412 23030 5 (Hardback)
ISBN 0 412 23040 2 (Science Paperback)

British Library Cataloguing in Publication Data

Barnard, J.A.
Flame and combustion.—2nd ed.
1. Combustion
I. Title II. Bradley, J.N.
541.3′61 QD516

ISBN 0–412–23030–5
ISBN 0–412–23040–2 Pbk

Library of Congress Cataloging in Publication Data

Barnard, J.A.
Flame and combustion.

Rev. ed. of: Flame and combustion phenomena / John N. Bradley. 1969.
Bibliography: p.
Includes index.
1. Combustion. 2. Flame. I. Bradley, John N. II. Bradley, John N. Flame and combustion phenomena. III. Title.
QD516.B29 1984 541.3′61 84–14247
ISBN 0–412–23030–5
ISBN 0–412–23040–2 (Science paperbacks; pbk.)

Contents

Preface to the first edition

The book is intended to serve as a primer to combustion. It has been the author's experience that too many scientists with interests in combustion phenomena have very limited knowledge of the field as a whole. For example, many chemists who have acquired a deep understanding of the mechanism of branching-chain reactions in closed vessels are completely uninformed about the importance of such processes in flames or detonation waves. This is a severe limitation because the essential feature of all combustion phenomena is that they arise as a result of the interplay of physical and chemical processes and a complete understanding can result only if aspects of mechanical engineering and fluid mechanics are taken into account. The aim of this text is to provide the basic principles which form the background to all combustion phenomena. It is based on a course given to postgraduate students in chemistry at the University of Essex and it is the author's hope that it can be read by final-year undergraduates and research personnel in a wide range of disciplines.

The major problem for the author has been that of selection. Because the book is intended to be short, many topics of interest have been omitted and, since decisions as to content have been entirely arbitrary, many readers will disagree with the choice. The author has tried to adhere to certain principles in making the selection. The most drastic has been to omit as far as possible any mention of experimental techniques and to minimize presentation of experimental evidence. The unfortunate consequence of this is that many arguments are presented as absolute and unequivocal when, in fact, they are supported merely by the balance of evidence, and individual investigators will possess quite diverse opinions. The author could not find any obvious way to avoid this difficulty and naturally hopes that the person who wishes to make further use of the material will adopt the suggestions made for further reading and hence obtain the benefit of the experimental background.

The layout of the book should be obvious from the contents page. The first chapter simply summarizes the various combustion phenomena and shows how they are related one to another. The following three chapters deal with the main combustion phenomena in gases emphasizing only the physical processes involved. The chemical processes which occur in gaseous combustion systems are discussed in Chapters 5, 6 and 7. Chapter 8 then describes how the combustion phenomena are modified when they occur in disperse media or condensed phases. The final four chapters deal with practical applications of these phenomena. Brief descriptions of the engineering aspects are given so that the demands on the combustion system may be better understood.

In these chapters, it has proved quite impossible to attempt any completeness of coverage, however broad, and the author has therefore selected items which appear to have some particular interest. For example, the chapter on rocket propulsion is largely devoted to a discussion of the thermochemistry of propellants because the considerations involved must bear on the whole range of combustion phenomena.

The treatment has been biased towards the needs of the physical scientist rather than the engineer. A basic knowledge of physics and chemistry is assumed although inevitably some aspects of the phenomena dealt with require more background information than others. Mathematical analysis has been limited as much as possible, the necessary expressions being introduced with a minimum of explanation. Emphasis has been placed throughout on the principles involved with only brief mention of numerical data or practical examples. To a large extent each chapter is self-contained so that cross-referencing has been kept to a minimum and some of the essential points have been repeated. References are only quoted where it is felt they are particularly relevant, for example, when a theory which explains some effect can be identified by the name of an individual investigator.

The author wishes to express his deepest gratitude to Miss P.J. Tweed, for typing the manuscript and for her assistance in its preparation, and to Professor P. Gray and Professor P.G. Ashmore, for their very valuable comments on, and contributions to, this book.

J.N.B.

Preface to the second edition

At the time of his sudden tragic death in February 1981, John Bradley had just started to revise his short book on combustion. Some months later I accepted an invitation from the publishers to complete this task.

Although the last decade has seen the publication of a great deal of new material on combustion, I have attempted to preserve the style of the original and not to increase the length of the book unduly. The basic plan of the original has been followed but where there have been significant advances these have been mentioned and some new material, particularly on combustion and the environment, has been incorporated. There is also a brief section on the fundamentals of thermodynamics and rate processes as they apply to combustion. This section is intended as a 'refresher' for those whose knowledge in these areas is a little rusty: the uninitiated seeking a more thorough treatment must consult appropriate texts where these subjects are covered in greater depth. Within the compass of a book of this length, it is quite impossible to cover all aspects of such a vast and many-sided field as combustion: inevitably some important topics are not covered in detail and some are not mentioned at all. More references to original work and specialist monographs are included, but in no sense is the bibliography intended to be comprehensive; rather the references and suggestions for further reading are intended as a first guide for those who wish to explore aspects of the subject in more detail.

John Bradley's intention was that his book should be suitable for final-year undergraduates and those beginning research in a wide range of disciplines. It is my hope that the same will be true of this new edition. With the needs of students particularly in mind, several chapters end with problems designed to illustrate important principles without at the same time requiring excessive mathematical manipulation. Too much attention should not be paid to the exact answers obtained; more accurate figures would be obtained by a more detailed, and hence lengthier, treatment.

Many colleagues have helped me in my task. Particular thanks are due to Professor D. Bradley, Professor C.F. Cullis, Professor P. Gray, Dr A.N. Hayhurst, Dr V.D. Long and Dr R.W. Walker for their invaluable comments and advice. For the errors which doubtless remain, I alone am responsible.

J.A.B.

Glossary

LIST OF SYMBOLS

A, A'	Pre-exponential term
$\boldsymbol{A}$	Area
B, B'	Constants
B	Transfer number
C	Molar heat capacity
D	Diffusion coefficient
$\boldsymbol{D}$	Detonation velocity
E	Activation energy
E'	Strain energy
E_{I}	Ionization potential
F	Thrust
$\boldsymbol{F}$	Fraction
G	Gibbs free energy
H	Enthalpy
I_0	Initiation rate
I_{s}	Specific impulse
$I_{\mathrm{s},\rho}$	Density specific impulse
J	Quantum number
J	Reaction rate in terms of mass per unit volume
J'	Reaction rate in Mallard–Le Chatelier model of a flame
K	Equilibrium constant
K_c	Equilibrium constant in terms of concentrations
K_p	Equilibrium constant in terms of partial pressures
L	Length scale for macroturbulence
Le	Lewis number
M	Molar mass
Ma	Mach number
P	Steric factor
Pe	Peclet number

Q	Exothermicity of reaction
R	Universal gas constant
R'	Gas constant per unit mass of gas
Ra	Rayleigh number
Re	Reynolds number
Re_L	Turbulent flow Reynolds number based on length scale L
Re_λ	Turbulent flow Reynolds number based on length scale λ
S	Entropy
S	Surface area
S_u	Laminar burning velocity
S_t	Turbulent burning velocity
Sh	Sherwood number
T	Absolute temperature
U	Internal energy
U_s	Shock velocity
V	Volume
W	Particle velocity
W	Burning rate
X	General reactant
Z	Collision frequency
a	Thermal diffusivity
a	Speed of sound
a, b, c, d	Constants
b	Co-volume
c	Velocity
c	Concentration
c	Specific heat capacity
c.r.	Compression ratio
d	Distance
d	Diameter
d_Q	Quenching distance between parallel plates
d_T	Quenching diameter in a tube
e	2.71828
e	Specific internal energy
e	Electron
f	Function of
f	Rate factor
g	Degeneracy
$\mathbf{g}$	Gravitational acceleration
g_B	Blow-off limit

g_F	Flash-back limit
h	Heat transfer coefficient
$\boldsymbol{h}$	Planck constant
h	Specific enthalpy
i, j	Numbers
k	Rate constant
$\boldsymbol{k}$	Boltzmann constant
k_g	Mass transfer coefficient
l	Mean free path
l	Heat required to vaporize a unit mass of fuel
m	Mass
$\dot{m}$	Mass flow rate
n	Number
$\dot{n}$	Molecular flow rate
p	Pressure
q	Heat release per unit mass
$\dot{q}_+$	Rate of heat generation
$\dot{q}_-$	Rate of heat loss
r	Radius
$\boldsymbol{r}$	Rate of reaction in moles per unit volume per unit time
s	Stoichiometric coefficient, moles of oxidant consumed per mole of fuel
t	Time
u	Velocity
v	Specific volume
v'	Volumetric flow rate
w	Mass fraction
x	Distance
y	Mole fraction
y	Height
x, y, z	Cartesian co-ordinates
α	Coefficient of thermal expansion
α	Area ratio
α_d	Degree of dissociation
β	Collision efficiency
β	Density ratio
β	Fraction
γ	Specific heat ratio
δ	Dimensionless rate of heat release
δ	Thickness

ε	Thermal efficiency
η	Dynamic viscosity
κ	Burning-rate coefficient
λ	Thermal conductivity
λ	Scale parameter
μ	Number
ν	Stoichiometric number
ν	Frequency
ν'	Quantum number
$\bar{\nu}$	Mean chain length
π	3.14159
ρ	Density
σ	Molecular diameter
τ	Lifetime
ϕ	Net branching factor
χ	Multiplication factor
ψ	Angle
ω	Probability of a branching collision
Γ	Specific volume ratio
Δ	Difference between two states
Π	Pressure ratio
Σ	Sum of
Φ	Equivalence ratio
∇^2	Laplacian operator
Δ'	Depth ratio

Subscripts and superscripts

B	Burner
C	Combustion chamber
E	Exhaust
F	Fuel
I	Ionization
INERT	inert
M	Mixture
O_2	Oxygen
OX	Oxidant
PR	Products
V	Constant volume

b	Branching
b	Burning
c	Concentration
c	Critical
d	Droplet
d	Dissociation
e	Electron
e	Equilibrium
f	Final, formation, forward, flame
fl	Fluctuation
g	Gas phase
i	Ignition
j	Identifier
l	Limit
l	Lobe
m	Minimum
o	Initial, inlet
opt	Optimum
p	Constant pressure
p	Propagation
p, s, t	Primary, secondary, tertiary
pz	Preheating zone
r	Reverse
rz	Reaction zone
s	Constant entropy
s	Surface
st	Stationary
t	Termination
total	Total
w	Wall
1	Region of undisturbed gas
2	Region behind shock or detonation front
$\ominus$	Standard state
*	Electronically excited state
†	Vibrationally excited state
+	Positive ion
−	Negative ion

A time derivative is denoted by a dot, thus, $\dot{n} = \mathrm{d}n/\mathrm{d}t$.

Useful constants

$\pi = 3.141\,59$

$e = 2.718\,28$

$\ln 10 = 2.302\,59$

Planck constant, $\boldsymbol{h} = 6.6262 \times 10^{-34}$ J s
Avogadro constant, $N = 6.022 \times 10^{23}$ mol^{-1}
Boltzmann constant, $\boldsymbol{k} = 1.380\,66 \times 10^{-23}$ J K^{-1}
Universal gas constant, $R = 8.3144$ J mol^{-1} K^{-1}
Mass of the electron, $m_e = 9.1095 \times 10^{-31}$ kg
Speed of light in a vacuum, $c = 2.9979 \times 10^8$ m s^{-1}
Standard gravitational acceleration, $g = 9.807$ m s^{-2}
Electron volt, eV $= 1.602 \times 10^{-19}$ J

Volume of 1 kmol of ideal gas at stp = 22.41 m^3
stp = $1.013\,25 \times 10^5$ N m^{-2} and 273.15 K

1 standard atmosphere = $1.013\,25 \times 10^5$ N m^{-2}
= 760 torr

1 torr = 133.32 N m^{-2}
Molar mass for air = 28.96 kg kmol^{-1}

Composition of air	N_2	O_2	Ar	CO_2
By volume	78.09%	20.95%	0.93%	0.03%
By weight	75.53%	23.14%	1.28%	0.05%

For approximate calculations the following may be used:

Atomic masses

H	$1\,kg\,kmol^{-1}$	N	$14\,kg\,kmol^{-1}$
C	$12\,kg\,kmol^{-1}$	S	$32\,kg\,kmol^{-1}$
O	$16\,kg\,kmol^{-1}$		

Composition of air	N_2	O_2
By volume	79%	21%
By weight	76.7%	23.3%

For air

$c_p = 1.005\,kJ\,kg^{-1}\,K^{-1}$

$c_V = 0.718\,kJ\,kg^{-1}\,K^{-1}$

$\gamma = 1.40$

1

Introduction

Combustion provided early man with his first practical source of energy: it gave him warmth and light, it extended the range of foodstuffs which he could digest, and it enabled him to 'work' metals. Throughout the world today, combustion still provides more than 95% of the energy consumed and, despite the continuing search for alternative energy sources, there is little doubt that combustion will remain important for centuries to come, particularly where a convenient method of storing energy is required, as for example in transport applications.

However, the reserves of fossil fuels are becoming depleted which means that the fuels themselves are becoming more expensive and less accessible sources have to be tapped. It has become increasingly important therefore to ensure that combustion is employed in the most efficient manner possible. At the same time almost all applications of combustion have adverse effects on the environment and it is essential to ensure that these are kept to an absolute minimum and are not forgotten in the search for improved efficiency. Thus the study of combustion processes is an important and growing area of science and this is reflected in the dramatic increase in published work during the last decade.

Combustion phenomena arise from the interaction of chemical and physical processes. The practical application of combustion involves not only physics and chemistry but also applied sciences such as aerodynamics and mechanical engineering. Because of the breadth of the subject, experimental investigations have concentrated on isolated phenomena, for example, flames, explosions etc., or on approaches based on only one of the sciences. The various textbooks covering the subject clearly mirror these attitudes. The aim of the present text is to bring together, at a sufficiently elementary level, the various facets and to show how the underlying processes are interrelated. In order to achieve the necessary concentration of information, much important material has been omitted, experimental evidence for almost all the statements made has been excluded, and occasionally rather bold generalizations have been necessary. The reader is asked to bear these points in mind in the knowledge that more advanced texts are available which deal with separate sections of the field and which are free from these short-comings.

Combustion commences in chemistry with the occurrence of a self-supporting exothermic reaction. Some of the phenomena, for example, the detonation wave, arise almost entirely from the chemical energy released, and it is only necessary to consider the chemistry of the system to the extent that it provides a source of energy at a rate which depends on the prevailing temperature, pressure and reactant concentrations. Other effects, such as the emission of light, depend on specific chemical processes which may have a quite negligible effect on the main phenomenon.

The physical processes involved in combustion are principally those which involve transport of matter and transport of energy. The conduction of heat, the diffusion of chemical species, and the bulk flow of gas all follow from the release of chemical energy in an exothermic reaction. It is the interaction of the various processes which leads to the phenomena observed. In the way that specific chemical reactions prove to be responsible for important 'side-effects', various physical processes may be similarly involved.

1.1 SURVEY OF COMBUSTION PHENOMENA

In combustion the chemical reaction normally involves two components, one of which is termed the *fuel* and the other the *oxidant*, because of the part which each plays in the reaction. Perhaps the simplest combustion system one can envisage arises when the premixed materials, in gaseous form, are heated slowly in a closed container. Provided the temperature does not rise above a certain value, the heat produced by the reaction will be dissipated at the vessel walls. A steady state is thus established and reaction proceeds smoothly to completion: this situation is referred to as *slow combustion*. This is principally of interest to the chemist since there is no strong interaction between any physical and chemical factors.

However, beyond a certain critical limit, which depends on the physical properties of both the reactants and the container, the rate of energy release by chemical reaction may exceed the rate at which it can be lost from the vessel by the various processes of heat transfer. If this occurs, the temperature rises and, in consequence, the rate of reaction and hence the rate of energy release both increase even further. The reaction rate thus accelerates indefinitely (so long as the supply of reactants is adequate) leading to an *explosion*. Strictly speaking, the term 'explosion' refers to the violent increase in pressure which must accompany the rapid self-acceleration of the reaction. The state at which self-acceleration occurs is termed ignition and the corresponding temperature the *ignition* or, more correctly, *self-ignition temperature*. Because the energy is released in the form of heat, this type of event is classified as a *thermal explosion*. Self-acceleration occurs because rates of reaction vary exponentially with temperature, whilst heat transfer by conduction depends linearly on temperature.

Energy may also be released chemically in the form of highly active intermediates, usually atoms or free radicals. These species can participate in *chain reactions*. In each step of a chain reaction a reactant molecule is consumed and a further active entity is generated. These chains are of finite length due to competing processes which remove active centres. A stable situation can therefore be established in which the rate of formation of active species is equal to their rate of removal, each active species formed being responsible for the consumption of a number of reactant molecules before it is itself destroyed. Chain reactions of this type thus comprise a step in which radicals are produced from stable molecules, this step being known as chain initiation, followed by propagation reactions in which a radical reacts with a reactant molecule resulting in the formation of products and another radical to continue the chain, and termination processes in which the chain-carrying radicals are removed. In some reactions one active centre reacts to give two or more further reactive species. The chain is then no longer *linear*, but is said to *branch*, and the rate of reaction is no longer constant but increases exponentially leading to a *branching-chain explosion*. This situation is formally similar to that which occurs in a nuclear explosion.

The distinction between these two mechanisms is important for an understanding of combustion phenomena and it will be referred to repeatedly in later chapters. It should be emphasized that, in any explosive system, both the concentration of active species and the temperature will normally display an exponential increase and it is frequently very difficult to decide which of the two processes is critical. Indeed, in many cases, a proper explanation demands the simultaneous involvement of both mechanisms.

The terms *degenerate branching*, *delayed branching* and *degenerate explosion* are used to describe a special type of branching-chain reaction in which the multiplication of chain centres occurs only very infrequently. The effect observed is again of an exponentially increasing reaction rate, but the build-up is far more gradual so that in a closed system the rate eventually falls off due to consumption of reactants before a violent explosion can occur.

In the discussion so far, explosion has been assumed to occur uniformly throughout the volume of the containing vessel. However, if ignition occurs in a localized region, a zone of explosive reaction can propagate through the reactant mixture. If the gases are contained in a long tube and reaction is initiated at one end, for example, by a spark, then a *combustion wave* will travel down the tube. Such a wave is also described as a *flame* or a *deflagration*, although the former term is usually restricted to a combustion wave in which the reaction zone is luminous. Some authors use the term explosion wave in this context but this invites confusion with a distinct phenomenon, *detonation*, which will be described later. Even when an attempt is made to achieve homogeneous initiation, ignition will usually commence locally and some form of combustion wave passes through the system.

If the reactant gas is forced to travel at the appropriate velocity towards the flame front, the flame should come to a standstill. In practice, it is normally necessary to construct a *burner* which holds the flame in one position and renders it stable towards small disturbances, such as variations in flow velocity. In a propagating combustion wave, reaction is induced in the layer of gas approaching the flame front and two possible mechanisms can be visualized which are analogous to the thermal and branched-chain mechanisms observed in static systems. In the former, *heat conduction* from the hot, burnt gas causes reactions to commence while, in the latter, *diffusion* of active intermediates from the reaction zone initiates combustion. Thus we see combustion phenomena becoming progressively more complex as diffusion, heat transfer and mass flow are able to interact with the chemistry.

In the situations examined so far, the reactant gases are intimately mixed and the flame is said to be *pre-mixed*. An alternative type of flame, known as a *diffusion flame*, arises when the two streams of reactant gases are initially separate and reaction takes place at the interface between them. A familiar example of a diffusion flame is provided by the candle flame. Here the heat from the flame causes evaporation of the wax which then acts as the fuel and burns as it mixes with the surrounding air.

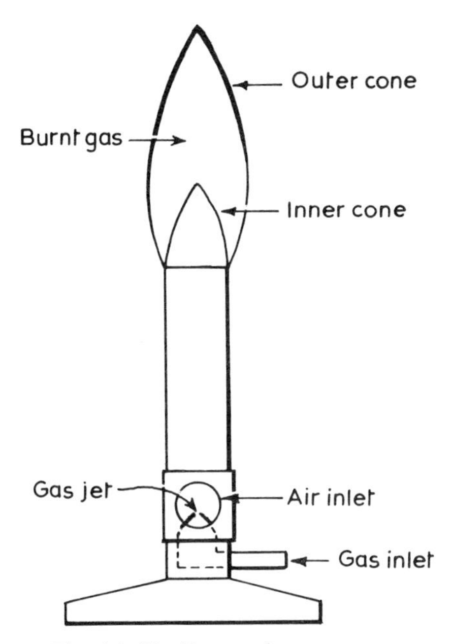

Fig. 1.1 The Bunsen burner.

The well-known *Bunsen burner* (Fig. 1.1) can be used to illustrate both types of flame. The fuel gas, entering through the nozzle at the base, entrains air, due to the Venturi effect, and the two gases mix as they travel up the tube. The *inner cone* is the reaction zone of a pre-mixed flame. Up to this stage, the mixture is fuel-rich so the composition of the burnt gas does not correspond to complete combustion. The *outer cone* is then a diffusion flame between the burnt gas and the surrounding air. Although the two flames coincide at the burner rim, they can be isolated in a picturesque manner on a *Smithells separator* [1] (Fig. 1.2).

In systems containing hydrocarbon fuels, a combustion wave is sometimes observed in which the explosive process is self-quenching. Reaction ceases when only a small fraction of the reactants has been converted to products and the temperature has risen by only some tens of degrees. This is termed a *cool flame*. In many combustion systems there is an appreciable interval between the act of initiation and the apparent onset of combustion or ignition. This is known as the *ignition delay* and cool flames are sometimes observed during this pre-ignition period.

The velocities of the types of combustion wave described so far are

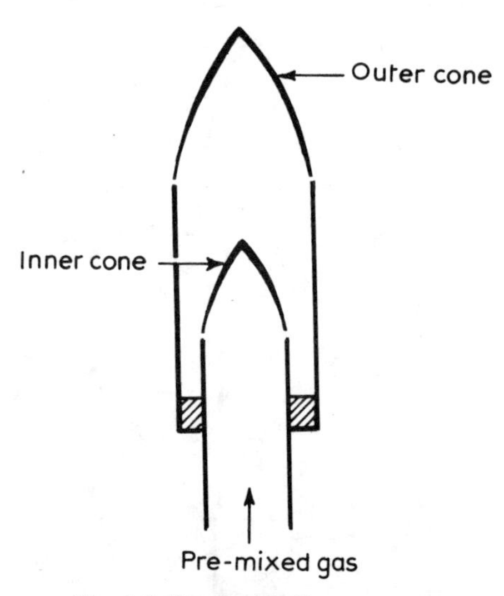

Fig. 1.2 The Smithells separator.

limited by transport processes, for example, heat conduction and diffusion, and cannot exceed the speed of sound in the reactant gas mixture. However, it is often found that a propagating combustion wave undergoes a transition to a quite different type of wave travelling at a much higher velocity, well above the sound speed. In this *detonation wave* reaction is initiated by a supersonic compression or *shock wave* travelling through the reactants. The energy released by chemical reaction behind the shock front provides the driving force for the shock wave.

The phenomena described above are not necessarily restricted to gaseous media. Indeed, most of them can occur equally well in both liquids and solids and also in dispersions of one phase in another (for example, mists and dust clouds) or at the interfaces between bulk phases.

The military, commercial and industrial applications of combustion are well known. The destructive power of explosives is used in weapons and in industrial situations such as blasting. In *high explosives*, where the maximum amount of energy is released in the minimum period of time, the physical process involved is characteristic of a detonation rather than a slower combustion wave. *Low-order explosives*, which find a particular application as propellants, are those in which deflagration occurs and energy is released much more slowly.

Combustion is used for heating in a range of applications varying from the domestic coal fire to the oil burner of a large industrial furnace. While, for reasons of economy, the oxidant is usually atmospheric air, the fuel may be solid, liquid or gas, and in any degree of dispersion.

The energy released in combustion may subsequently be used as a source of motive power as, for example, when it is used to heat the boiler of a steam engine. However, it is more usual to transform the heat energy into mechanical energy within the engine itself, hence the name *internal combustion engine.* The actual sequence of events leading from the release of chemical energy to the production of mechanical power differs depending on whether the device is a reciprocating piston engine, a gas turbine, a jet engine or a rocket motor. A chemical fuel provides a very convenient way of storing energy in an easily transportable form so that combustion is especially valuable in propulsion, the vehicle carrying its own self-contained energy source as well as the required payload. Indeed, rockets and modern jet engines can even support the total weight of the vehicle for a substantial period without making use of aerodynamic lift.

1.2 THERMODYNAMICS OF COMBUSTION

Combustion involves the liberation of energy by chemical reaction. Matters related to the amount of energy released by reaction under given circumstances and the equilibrium composition of a set of reactants all fall within the scope of thermodynamics. While a full treatment of this subject would not be appropriate in the present volume, and indeed it is not unreasonable to assume that readers already possess some knowledge of thermodynamics, it nevertheless seems worthwhile to review briefly those thermodynamic principles which are particularly important in combustion.

The first requirement is to know how much energy is available from a given chemical reaction. In order to obtain this information we make use of the energies of the individual reactants and products. The precise products, and hence the overall stoichiometry of the reaction, must always be established. For example, methane can react with oxygen in two ways, producing either carbon monoxide

$$CH_4 + \tfrac{3}{2}O_2 \rightarrow CO + 2H_2O \tag{1.1}$$

or carbon dioxide

$$CH_4 + 2O_2 \rightarrow CO_2 + 2H_2O \tag{1.2}$$

While the second reaction corresponds to complete combustion, and hence to a greater release of energy, there are many circumstances in which the first reaction provides a better representation of what actually happens when methane is burned.

The energy of an individual chemical species may be given either in terms of its *internal energy*, U, or *its enthalpy*, H, where, by definition, $H = U + pV$. While these quantities are quite easily interconvertible, it is more usual to tabulate enthalpy values. Since neither internal energy nor enthalpy can be measured absolutely, it is necessary to choose a reference state to which all other energies may be related: this standard state is generally taken as the stable state of the pure substance at atmospheric pressure and under these conditions the enthalpy of each element is arbitrarily assigned the value zero. The enthalpy change involved at some temperature T in forming one mole of the chemical species in its standard state from its elements in their standard state is known as the standard enthalpy of formation, $\Delta_f H_T^{\ominus}$, of the compound. There are extensive tabulations of these standard enthalpies [2–5]

Table 1.1 Some standard enthalpies of formation (all species are gaseous, unless otherwise stated).

Species	Formula	$\Delta_f H_{298}^{\ominus}$ (kJ mol^{-1})
Hydrogen atom	H	217.99
Oxygen atom	O	249.19
Hydroxyl radical	OH	39.46
Hydroperoxy radical	HO_2	20.92
Water liquid	H_2O (1)	−285.83
Water vapour	H_2O	−241.81
Hydrogen peroxide liquid	H_2O_2(1)	−136.1
Carbon monoxide	CO	−110.52
Carbon dioxide	CO_2	−393.51
Methyl radical	CH_3	145.7
Methane	CH_4	− 74.87
Ethyne (acetylene)	C_2H_2	226.7
Ethene	C_2H_4	52.47
Ethane	C_2H_6	− 84.67
Propane	C_3H_8	−103.85
Benzene	C_6H_6	82.93
Methanol vapour	CH_3OH	−201.2
Nitrogen atom	N	472.7
Amine radical	NH_2	167.7
Ammonia	NH_3	− 45.94
Nitric oxide	NO	90.29
Sulphur dioxide	SO_2	−296.81

usually given at 25° C (298 K). A selection of data of particular relevance to combustion is given in Table 1.1.

We are now in a position to calculate the energy available from a typical chemical reaction. From Hess's Law (which is a special case of the first law of thermodynamics), the enthalpy change in the reaction taking place under standard state conditions $\Delta H_T^\ominus$, is equal to the difference between the sums of the standard enthalpies of formation of the products and the reactants. For Reaction 1.1 when the water is produced as vapour

$$\begin{aligned}\Delta H_{298}^\ominus &= [-110.52 + 2(-241.81)] - [-74.87 + (3/2) \times 0] \\ &= -519.27\,\text{kJ mol}^{-1}\end{aligned} \tag{1.3}$$

the negative sign indicating that energy is released. Such a reaction is called *exothermic* and combustion reactions are, of course, in this category. A reaction for which ΔH is positive and in which, therefore, energy is absorbed by the system from the surroundings, is said to be *endothermic*.

For the general reaction

$$\nu_A A + \nu_B B + \cdots \rightarrow \nu_P P + \nu_Q Q + \cdots \tag{1.4}$$

$$\Delta H_{298} = \sum \nu_X \Delta_f H_{298}^\ominus(X) \tag{1.5}$$

where the summation is performed treating stoichiometric numbers, ν_X, as positive for the products and negative for the reactants.

It is essential to realize that ΔH for a chemical reaction can only be quoted unambiguously if the stoichiometric equation is also given. The units of ΔH are J mol^{-1}, that is, joules for an extent of reaction of one mole. The concept of extent of reaction is a somewhat subtle one and is discussed fully elsewhere [6]. For the present purpose, it is sufficient to remark that any stoichiometric equation only expresses the combining ratios of atoms and molecules: it says nothing about the scale on which the reaction is carried out. When, for example, we write

$$2CO + O_2 \rightarrow 2CO_2 \tag{1.6}$$

this is short-hand for 'when carbon monoxide combines with oxygen to form carbon dioxide, the ratio of the number of molecules of the three substances involved is 2:1:2'. When we add $\Delta H_{298}^\ominus = -566\,\text{kJ mol}^{-1}$ we are now saying that 'when two moles of carbon monoxide combine with one mole of oxygen to form two moles of carbon dioxide, the enthalpy change (under standard conditions at 298 K) is -566 kJ'.

For some engineering purposes it is sometimes preferable to use the enthalpy released per unit mass of reactants: for Reaction 1.1 this is $519.27/(16 + \frac{3}{2} \times 32) = 8.11\,\mathrm{kJ\,g^{-1}}$. As atmospheric air is usually regarded as 'free', another useful quantity is the amount of energy available from unit mass of fuel; for methane burning to carbon monoxide, the energy available is $519.27/16 = 32.96\,\mathrm{kJ\,g^{-1}}$.

These calculations relate the enthalpy changes taking place at constant pressure. The corresponding energy released when reaction takes place at constant volume is given by the internal energy change, ΔU. For reactions involving gases there is a simple relationship between ΔH and ΔU.

$$\Delta H = \Delta U + (\Delta \nu)RT \tag{1.7}$$

where $\Delta \nu$ is the change in the number of moles of gaseous reactants. When only liquids or solids are involved, in most cases, $\Delta H \simeq \Delta U$.

Combustion reactions do not usually take place at room temperature but the enthalpy change at other temperatures is easily obtained from

$$\Delta H_T = \Delta H_{298} + \int_{298}^{T} \Delta C_p \,\mathrm{d}T \tag{1.8}$$

where ΔC_p refers to the difference between the heat capacity of the reactants and that of the products, that is,

$$\Delta C_p = \sum \nu_{\mathrm{X}} C_p(\mathrm{X}) \tag{1.9}$$

where the summation is performed, as before, treating the stoichiometric numbers as positive for the products and negative for the reactants. $C_p(\mathrm{X})$ represents the molar heat capacity of reactant X. In practice, the enthalpy changes are not very sensitive to temperature because the heat capacities on each side of the chemical equation are roughly equal. For Reaction 1.1, already considered. ΔH only changes from $-519\,\mathrm{kJ\,mol^{-1}}$ at 298 K to $-530\,\mathrm{kJ\,mol^{-1}}$ at 2000 K and $-542\,\mathrm{kJ\,mol^{-1}}$ at 3000 K.

In real systems, the energy is released over a range of temperatures. Fortunately, both internal energy and enthalpy are state properties, which means that their value depends only on the present state of the system and not the path by which it was reached. It follows that, so long as the initial and final states of the system are fixed, we can calculate the change in internal energy or enthalpy associated with the process, and this change will be independent of the route by which the change takes place. Thus, we can find the enthalpy change for a system which commences with reactants at an initial temperature T_0 and finishes with products at a final temperature T_f by first finding the enthalpy change at T_0 and then using heat capacity data to calculate the enthalpy change involved in heating the products from T_0 to T_f.

This leads naturally into the calculation of *adiabatic reaction temperatures*. For a closed system, the first law of thermodynamics tells us that the energy released in a reaction goes into raising the temperature of the system.

For a *constant volume process*, such as a closed vessel explosion, this means that

$$-\Delta U_{T_0} = \int_{T_0}^{T_f} C_V\,(\text{products})\,\mathrm{d}T \tag{1.10}$$

and similarly for a *constant pressure process*, typified by a pre-mixed flame

$$-\Delta H_{T_0} = \int_{T_0}^{T_f} C_p\,(\text{products})\,\mathrm{d}T \tag{1.11}$$

Heat capacities are to be found in compilations of thermodynamic data either tabulated against temperature [4, 5] or, more conveniently as

Table 1.2 Molar heat capacities of some gases.
$C_p = a + bT + cT^2 + dT^3$ where C_p is the molar heat capacity in J mol^{-1}K^{-1} and T is the temperature in Kelvins.

Species	a	$10^2 b$	$10^5 c$	$10^9 d$	Valid over temperature range (K)
H_2	29.09	−0.1916	0.4000	−0.870	273–1800
O_2	25.46	1.519	−0.7150	1.311	273–1800
N_2	27.32	0.6226	−0.0950	—	273–3800
CO	28.14	0.1674	0.5368	−2.220	273–1800
CO_2	22.24	5.977	−3.499	7.464	273–1800
H_2O	32.22	0.1920	1.054	−3.594	273–1800
CH_4	19.87	5.021	1.286	−11.00	273–1500
C_2H_2	21.80	9.208	−6.523	18.20	273–1500
C_2H_4	3.95	15.63	−8.339	17.66	273–1500
C_2H_6	6.895	17.25	6.402	7.280	273–1500
C_3H_8	−4.042	30.46	−15.71	31.71	273–1500
C_6H_6	−39.19	48.44	−31.55	77.57	273–1500
CH_3OH	19.04	9.146	−1.218	−8.033	273–1000
NH_3	27.55	2.563	0.9900	−6.686	273–1500
NO	27.03	0.9866	0.3223	0.3652	273–3800
SO_2	25.76	5.791	−3.809	8.606	273–1800

polynomials in temperature [7, 8]

$$C_p = a + bT + cT^2 + dT^3 \tag{1.12}$$

Table 1.2 gives data relating to the heat capacities of some species involved in combustion.

For approximate calculations we can make use of the fact that the maximum value of C_V is given by the sum of the classical contributions, i.e.

Translational motion	$\frac{3}{2}R$
Rotational motion	$\frac{3}{2}R$ for a non-linear molecule
	R for a linear molecule
Vibrational motion	R for each mode

(There are $3n-5$ modes for a linear molecule and $3n-6$ modes for a non-linear molecule containing n atoms).

Values of C_p may be then obtained from the relationship

$$C_p = C_V + R \tag{1.13}$$

A mean heat capacity, suitable for rough estimates of temperatures, is obtained by assuming one-half of the classical vibrational modes to be active.

Particularly at the high temperatures involved in combustion processes there is a further factor which has to be taken into account. Chemical reactions never go to completion and the second law of thermodynamics provides criteria for the position of equilibrium and, in the case of a system at constant temperature and pressure, this may be represented by a minimum in the free energy change for the reaction, ΔG, where

$$\Delta G = \Delta H - T\Delta S \tag{1.14}$$

If the reactants and products behave as ideal gases (a reasonable approximation for most combustion systems), then the change in the Gibbs free energy accompanying reaction at a temperature T is related to the equilibrium constant K_p by the expression

$$-\Delta G_T^\ominus = RT \ln K_p \tag{1.15}$$

For the generalized Reaction 1.4

$$K_p = p_P^{\nu_P} p_Q^{\nu_Q} \cdots / p_A^{\nu_A} p_B^{\nu_B} \cdots \tag{1.16}$$

where p_A is the partial pressure of A etc. Defined in this way the equilibrium constant has dimensions of [pressure]$^{\Delta\nu}$. Just as there are tabulations of standard enthalpies of formation, there are lists of standard Gibbs free energies of formation for a very large number of compounds [2–5]; $\Delta G_T^\ominus$ for a particular reaction may be obtained by combining the tabulated $\Delta_f G_T^\ominus$ values in a way exactly analogous to the way $\Delta_f H_T^\ominus$ values are combined in Equation 1.5.

Equilibrium constants are functions of temperature and, if ΔH for the reaction is constant then

$$\ln(K_{p,T_2}/K_{p,T_1}) = \left(\frac{-\Delta H}{R}\right)\left(\frac{1}{T_2} - \frac{1}{T_1}\right) \tag{1.17}$$

If the second term in Equation 1.14 could be neglected, then the condition for equilibrium would correspond to a maximum in $-\Delta H$, the amount of heat evolved by the process: in other words, the reaction would always go to completion. However, the presence of the second

Table 1.3 Complete thermodynamic calculations for a stoichiometric propane–air mixture (quantities expressed in mole fractions).

Species	Constant pressure			Constant volume
	(a)	(b)	(c)	
CO_2	0.1004	0.1003	0.1111	0.0914
H_2O	0.1423	0.1439	0.1481	0.1374
N_2	0.7341	0.7347	0.7407	0.7276
CO	0.0099	0.0100		0.0182
H_2	0.0032	0.0033		0.0053
O_2	0.0048	0.0055		0.0075
NO	0.0020	0.0022		0.0052
CH_2O	$< 10^{-5}$			$< 10^{-5}$
C_2H_4	$< 10^{-5}$			$< 10^{-5}$
C_3H_6	$< 10^{-5}$			$< 10^{-5}$
N_2O	$< 10^{-5}$			$< 10^{-5}$
CHO	$< 10^{-5}$			$< 10^{-5}$
CH_3	$< 10^{-5}$			$< 10^{-5}$
C_2H_5	$< 10^{-5}$			$< 10^{-5}$
i-C_3H_7	$< 10^{-5}$			$< 10^{-5}$
H	0.0035			0.0009
O	0.0020			0.0008
N	$< 10^{-5}$			$< 10^{-5}$
OH	0.0027			0.0058
HO_2	$< 10^{-5}$			$< 10^{-5}$
Final temperature	2219 K	2232 K	2324 K	2587 K

Column (a) refers to calculations based on a comprehensive set of equilibria, column (b) refers to calculations using a smaller set of equilibria, while the figures in column (c) were obtained by considering only the formation of CO_2 and H_2O.

term means that a small reduction in $-\Delta H$ may be compensated for by a corresponding change in $T\Delta S$. The entropy, S, is a measure of disorder in the system; thus the entropy is increased by an increase in the number of different chemical species present, particularly if these are simpler, that is, contain fewer atoms. In a flame where methane burns to carbon dioxide and water, we may expect in addition to find traces of the following in the combustion products:

1. Unreacted material, CH_4 and O_2.
2. Other molecular products, e.g. CO and H_2.
3. Radical intermediates, e.g. H, O, OH and CH_3.

In general, when any reaction has reached equilibrium there will be varying amounts of chemical species other than the expected final products. At low temperatures, this complication is relatively unimportant in most combustion processes, but the presence of the temperature multiplier in the $T\Delta S$ term means that the trace materials become more important at elevated temperatures. The computer calculations in Tables 1.3 and 1.4 show that the effect is fairly small up to about 2000 K but has a dramatic effect at 3000 K. As the tables show, these temperatures tend to be associated with combustion in air and in oxygen, respectively.

For a system containing more than a handful of chemical species, establishing equilibrium conditions requires the use of a computer and the results in Tables 1.3 and 1.4 were obtained in this way.

However, an alternative approach is to look at the individual chemical equilibria involved. Although a condition of equilibrium is that all the possible equilibrium relationships are satisfied, in practice many of them will be interdependent and it is therefore necessary to consider only an appropriate subset. A good representation of the final situation resulting from the combustion of a hydrocarbon in air is provided by the following seven equilibria containing eleven chemical species:

The water-gas equilibrium:

$$CO + H_2O \rightleftharpoons CO_2 + H_2 \tag{1.18}$$

This largely controls the amount of carbon monoxide released during combustion.

The formation of nitric oxide:

$$N_2 + O_2 \rightleftharpoons 2NO \tag{1.19}$$

This equilibrium influences the yield of NO_x in engine exhausts.

Table 1.4 Complete thermodynamic calculations for a stoichiometric propane–oxygen mixture (quantities expressed in mole fractions).

Species	Constant pressure (a)	Constant pressure (b)	Constant pressure (c)	Constant volume
CO_2	0.1396	0.0896	0.4286	0.1293
H_2O	0.3000	0.3182	0.5714	0.2908
CO	0.1951	0.2544		0.2035
H_2	0.0697	0.1404		0.0693
O_2	0.0984	0.1974		0.0954
CH_2O	$< 10^{-5}$			$< 10^{-5}$
C_2H_4	$< 10^{-5}$			$< 10^{-5}$
C_3H_6	$< 10^{-5}$			$< 10^{-5}$
CHO	$< 10^{-5}$			$< 10^{-5}$
CH_3	$< 10^{-5}$			$< 10^{-5}$
C_2H_5	$< 10^{-5}$			$< 10^{-5}$
i-C_3H_7	$< 10^{-5}$			$< 10^{-5}$
H	0.0526			0.0460
O	0.0441			0.0443
OH	0.1005			0.1212
HO_2	3×10^{-5}			0.0001
Final temperature	3081 K	3420 K	6340 K	3630 K

Column (a) refers to calculations based on a comprehensive set of equilibria, column (b) refers to calculations using a smaller set of equilibria, while the figures in column (c) were obtained by considering only the formation of CO_2 and H_2O.

The dissociation equilibria:

$$2H_2O \rightleftharpoons 2H_2 + O_2 \tag{1.20}$$

$$H_2 \rightleftharpoons 2H \tag{1.21}$$

$$O_2 \rightleftharpoons 2O \tag{1.22}$$

$$N_2 \rightleftharpoons 2N \tag{1.23}$$

$$H_2O \rightleftharpoons H + OH \tag{1.24}$$

These equilibria have a considerable effect on the total amount of energy released in combustion.

Manual calculations of the equilibrium composition of the products of combustion [9] are extremely tedious and computer methods are now usually employed [10, 11].

1.3 RATE PROCESSES IN COMBUSTION

1.3.1 Basic reaction kinetics

While all combustion processes depend on the total amount of energy released by chemical reaction, not all depend on the rate of reaction, provided it exceeds some minimum value. Thus the gross behaviour of detonation waves, diffusion flames, burning droplets and liquid propellant rocket engines is virtually independent of chemical kinetics whilst pre-mixed flames, fires and internal combustion engines are all sensitive to the detailed kinetics involved. In this section the fundamentals of reaction kinetics are described: more details of specific reactions and reaction mechanisms appear in later chapters.

The quantitative behaviour of a chemical reaction is described by a *rate law* which specifies the time rate of change of the concentration of chemical species in terms of a product of concentration terms and a *rate constant* (or *rate coefficient*) which is independent of concentration but not usually of temperature. For the reaction represented by the stoichiometric equation

$$\nu_A A + \nu_B B + \cdots \rightarrow \nu_P P + \nu_Q Q + \cdots \tag{1.4}$$

the rate law takes the form

$$-\frac{1}{\nu_A}\frac{d[A]}{dt} = -\frac{1}{\nu_B}\frac{d[B]}{dt} = +\frac{1}{\nu_P}\frac{d[P]}{dt} = +\frac{1}{\nu_Q}\frac{d[Q]}{dt}$$

$$= k[A]^i[B]^j \tag{1.25}$$

where k is the rate constant. The powers i and j are known as the *order of reaction* with respect to A and B, respectively, and the overall order is $(i+j)$.

We must make a clear distinction between an *overall reaction*, often referred to as a *global reaction*, and an *elementary step*. In the latter, the chemical equation represents what actually happens on a molecular level. For example, the process

$$H + O_2 \rightarrow OH + O \tag{1.26}$$

is an elementary step; a hydrogen atom collides with an oxygen molecule and the three atoms rearrange to give a hydroxyl radical and an oxygen atom. In this case, the rate law may be written

$$-\frac{d[H]}{dt} = -\frac{d[O_2]}{dt} = +\frac{d[OH]}{dt} = +\frac{d[O]}{dt} = k[H][O_2] \tag{1.27}$$

the rate of reaction being directly proportional to the concentration of hydrogen atoms and oxygen molecules.

The overall equation, while it represents the stoichiometry of the reaction, does not necessarily reflect the detailed events as the species react. Indeed, many reactions take place by a series of elementary steps. Thus, the process

$$2H_2 + O_2 \rightarrow 2H_2O \tag{1.28}$$

may, under certain restricted circumstances, be described by the rate law

$$-\frac{1}{2}\frac{d[H_2]}{dt} = -\frac{d[O_2]}{dt} = +\frac{1}{2}\frac{d[H_2O]}{dt} = k[H_2]^{1.5}[O_2]^{0.7} \tag{1.29}$$

but in this case we must not think of two hydrogen molecules colliding with a single oxygen molecule to form two molecules of water; the global process is an abbreviation for a complex sequence of elementary steps. While the rate law for the elementary step is valid over a very wide range of temperature and pressure, the global rate law is an empirical representation of the dependence of the reaction rate on concentration and only applies over a very limited range of conditions. For a global reaction, the order has to be established by experiment; but if a reaction is known to be elementary, its order is immediately known without recourse to experiment. Ideally we would dispense with global reaction kinetics completely: however, it often happens that the detailed reaction mechanism is not completely understood, or that its complexity makes it unsuitable for incorporation into a computer model, and then a global expression has to be used.

An elementary step is said to be *unimolecular* if it involves only a single molecule, for example,

$$H_2 \rightarrow 2H \tag{1.30}$$

It then obeys *first-order kinetics* as expressed by the relationship

$$-\frac{d[H_2]}{dt} = +\frac{1}{2}\frac{d[H]}{dt} = k[H_2] \tag{1.31}$$

The *molecularity* of a reaction describes the number of molecules involved in the reactive event, while the *order* of a reaction is the exponent of the concentration term in the corresponding rate law. While unimolecular reactions are normally first order (except at very low pressures), a first-order reaction is *not* necessarily unimolecular.

A first-order rate constant has the dimensions of $[\text{time}]^{-1}$ and the units of time are almost always seconds. The relationship between the rate constant and temperature is expressed by the *Arrhenius equation*

$$k = A\exp(-E/RT) \tag{1.32}$$

For a particular reaction, the two *Arrhenius parameters*, A and E are constants and independent of temperature. A is termed the *frequency factor* or *pre-exponential term* and E is the *activation energy*. The frequency factor is related to the rate at which chemical bonds can rearrange in a molecule and hence is of the order of a vibrational frequency, that is, $10^{13}\,\text{s}^{-1}$. The activation energy is a measure of the energy barrier to reaction and therefore must be at least as great as the endothermicity.

In the gas phase, the energy needed to overcome the barrier is provided by binary collisions. At high pressures, the frequency of energizing collisions will balance the frequency of deactivating collisions and so the fraction of molecules with sufficient energy to react will be a constant, independent of pressure. At low pressures, the collision frequency will be inadequate to maintain the proportion of energized molecules. In consequence, the measured rate constant will display *fall-off* and the behaviour will approach that characteristic of second-order kinetics. This provides an example of a situation in which the molecularity and the order of reaction differ. The theory of unimolecular reactions is well-developed and both frequency factors and fall-off behaviour can often be predicted with a precision satisfactory for most practical purposes [12–14].

The bimolecular reaction

$$H + O_2 \rightarrow OH + O \tag{1.33}$$

displays *second-order kinetics*

$$-\frac{d[H]}{dt} = -\frac{d[O_2]}{dt} = +\frac{d[OH]}{dt} = +\frac{d[O]}{dt} = k[H][O_2] \tag{1.34}$$

since the sum of the exponents of the concentration terms is two. The rate constant now has dimensions of $[\text{concentration}]^{-1}\,[\text{time}]^{-1}$; the temperature dependence of k may be expressed in Arrhenius form, as before. An elementary treatment of bimolecular reactions leads to the notion that the Arrhenius pre-exponential factor A may be regarded as the product of the *collision frequency* of the reactants, Z, and a *steric factor*, P, which reflects the importance of the orientation of the reactants at the moment of collision. The activation energy, E, is provided by the relative kinetic energy of the collision partners along the line-of-centres. More sophisticated theories of bimolecular reaction rates are available and these lead to expressions of the form

$$k = A'T^n \exp(-E/RT) \tag{1.35}$$

where n is a number of order unity, but the simple Arrhenius expression is sufficient for most purposes in the study of combustion.

The *termolecular* reaction

$$H + O_2 + M \rightarrow HO_2 + M \tag{1.36}$$

obeys *third-order kinetics*

$$-\frac{d[H]}{dt} \text{ etc.} = k[H][O_2][M] \tag{1.37}$$

In this example the symbol M is used to denote any molecule present in the system and its function is to remove some of the energy released by the formation of the new chemical bond thereby preventing the product from immediately redissociating. M is termed a *third body* or *chaperon*. The rate constant now has dimensions of [concentration]$^{-2}$ [time]$^{-1}$. Third-order reactions often have rates which fall slightly with increasing temperature and an Arrhenius representation leads to a small negative activation energy. In this case, the 'activation energy' has no physical significance and the use of an equation of Arrhenius form to represent the data is purely a matter of convenience. *Recombination* reactions of the type shown above are important in combustion because they are usually responsible for the release of much of the reaction energy.

It is useful to note that the forward and reverse rate constants of an elementary reaction, k_f and k_r, can be related through the expression $k_f/k_r = K_c$, where K_c is the equilibrium constant defined in terms of concentration.

Combustion reactions normally involve a complex mechanism, or sequence of elementary steps, and many examples will be discussed in later chapters. Virtually all complex mechanisms involve *chain reactions*, that is, reactions in which an active species (usually a free radical or an atom) reacts with a stable molecule to give a product molecule and another active species which can propagate the chain. Thus, in the reaction between hydrogen and chlorine, the chain is propagated by the cycle:

$$Cl + H_2 \rightarrow HCl + H \tag{1.38}$$

$$H + Cl_2 \rightarrow HCl + Cl \tag{1.39}$$

Such reactions comprise a *linear chain* because each propagation step leaves the total number of active centres unchanged.

Although linear-chain propagation is normal, examples are known where *chain branching* occurs, that is, one active centre reacts producing more than one new centre capable of continuing the chain. In the hydrogen–oxygen system, the reaction

$$H + O_2 \rightarrow OH + O \tag{1.40}$$

brings about chain branching since both OH and O can react with hydrogen molecules to continue the chain. Branching chain reactions are particularly important in many combustion reactions and their special features will be discussed in Chapter 2, as well as being mentioned elsewhere later in the book.

Typical active species involved in hydrocarbon combustion are H, O, OH, CH_3 and CHO. Reactions of such species have low energy barriers and hence their rates are rapid even though the concentrations of active species are low. In general, reactions involving solely molecules are too slow to sustain combustion.

The behaviour of a complex reaction mechanism is described by a set of simultaneous differential equations, equal in number to that of the chemical species involved. Analytical solution of these equations is usually impossible and use is therefore made of the *stationary-state approximation* in which the derivatives of the concentrations of the active centres with respect to time are set equal to zero. In the hydrogen–chlorine example quoted above, one would write

$$\frac{d[H]}{dt} = \frac{d[Cl]}{dt} = 0 \tag{1.41}$$

This immediately reduces a number of differential equations to algebraic equations. This does not imply that the concentration of radicals does not change with time, but only that they are algebraically related to the concentrations of stable species, which have finite time derivatives. The algebraic equations can normally be solved to give the stationary-state concentration of radical species and thence the rate of attack on the reactant by the radicals can be obtained.

In some cases, when the rate of reaction is very high, it is not possible to use the stationary-state assumption, and then recourse must be made to numerical integration by computer.

1.3.2 Transport properties of gases

In looking at the physics of combustion we shall make considerable use of the laws of conservation of mass, momentum and energy. In some cases, we shall be concerned with the transport of these quantities through a fluid medium, that is to say with diffusion, viscosity and heat conduction, respectively.

The conduction of heat is described by *Fourier's Law*

$$\frac{\dot{q}}{A} = -\lambda \frac{dT}{dx} \tag{1.42}$$

where $\dot{q}$ is the heat flowing in unit time through an area A and λ is the thermal conductivity, which is measured in $J\,m^{-1}\,s^{-1}\,K^{-1}$ or $W\,m^{-1}\,K^{-1}$. The temperature gradient in the medium is dT/dx in the direction of heat flow and the negative sign indicates that the temperature gradient is in the opposite direction to the flow of heat. The thermal diffusivity, a, is defined as $\lambda/\rho c_p$ and thus has units of $m^2\,s^{-1}$.

For a gas, an expression for the thermal conductivity can be obtained using simple kinetic theory as follows.

Consider two planes A and B of unit cross-sectional area perpendicular to the temperature gradient and separated by a distance equal to the mean free path, l, that is, the average distance between molecules. Any molecule in the plane A at a temperature T_A travelling from left to right will retain its energy until it collides with another molecule which, on average, must occur at plane B. If the mean molecular velocity is $\bar{c}$, then half of the molecules crossing plane A at this velocity will do so from left to right. If the density of the gas is ρ and the heat capacity per unit mass at constant volume is c_V, the flow of heat from A to B in unit time will be $\frac{1}{2}\rho c_V \bar{c} T_A$. However, there will be a corresponding flow of heat from B to A of $\frac{1}{2}\rho c_V \bar{c} T_B$. Hence, the net heat flux from left to right is $\frac{1}{2}\rho c_V \bar{c}(T_A - T_B)$. But the temperature difference $(T_A - T_B)$ is equal to $-l\,dT/dx$ so that the net heat flux through unit area is $-\frac{1}{2}\rho c_V \bar{c} l\,dT/dx$.

Comparison with Fourier's Law shows that

$$\lambda = \tfrac{1}{2}\rho c_V \bar{c} l \tag{1.43}$$

This derivation is much oversimplified. More rigorous treatments result in a similar relationship but with a different numerical factor: all of them lead to an expression close to

$$\lambda = \tfrac{1}{3}\rho c_V \bar{c} l \tag{1.44}$$

The mean free path is inversely proportional to the density and $\bar{c}$ is proportional to $T^{1/2}$, so λ is virtually independent of pressure but increases with temperature very roughly as $T^{1/2}$. The measured value for air at room temperature is about 0.025 $W\,m^{-1}\,K^{-1}$ and hence the thermal diffusivity is about $2 \times 10^{-5}\,m^2\,s^{-1}$.

Diffusion is described by *Fick's Law*

$$\frac{\dot{n}}{A} = -D\frac{\mathrm{d}n}{\mathrm{d}x} \tag{1.45}$$

where $\dot{n}$ is the number of molecules per unit time crossing an area A, $\mathrm{d}n/\mathrm{d}x$ is the concentration gradient, and D is the diffusion coefficient with the same dimensions as those of thermal diffusivity. A simple kinetic theory derivation analogous to that given above predicts a *self-diffusion coefficient* equal to $\frac{1}{3}\bar{c}l$. The diffusion coefficient D is inversely proportional to pressure and increases with temperature according to about $T^{1.5}$. The measured value of D for a gas molecule such as nitrogen diffusing in air is around $2 \times 10^{-5}\,\mathrm{m^2 s^{-1}}$. It should be noted that the expression for the diffusion coefficient comprises the product of a velocity and a scale factor. When we come to describe transport due to turbulent or eddy motion later we shall adopt the same two parameters but with values characteristic of the eddies rather than the molecular motions.

The transport of momentum, or viscosity, is only involved indirectly in combustion, for example, when it controls the flow velocity profile in a tube.

The numerical values for the thermal diffusivity and the diffusion coefficient of the same molecules are very similar and the dimensionless ratio a/D known as the *Lewis number*, *Le*, is close to unity. It represents the ratio of conductive to diffusive fluxes and theoretical solutions to many problems in combustion are obtained by setting this number equal to one: this necessarily implies that both a and D show the same dependence on temperature.

1.4 SUGGESTIONS FOR FURTHER READING

Bett, K.E., Rowlinson, J.S. and Saville, G. (1975) *Thermodynamics for Chemical Engineers*, Athlone Press, London.

Laidler, K.J. (1965) *Chemical Kinetics*, 2nd edn, McGraw Hill, New York.

Mulcahy, M.F.R. (1973) *Gas Kinetics*, Nelson, London.

Warn, J.R.W. (1969) *Concise Chemical Thermodynamics*, Van Nostrand Reinhold, London.

Welty, J.R., Wilson, R.E. and Wicks, C.E. (1976) *Fundamentals of Momentum, Heat and Mass Transfer*, 2nd Edn, John Wiley, New York.

1.5 PROBLEMS

1. (a) Use the data in Table 1.1 to compare the maximum thermal energy available from hydrogen, methane, ethane, propane, ethine (acetylene), ethene and methanol vapour burning in air. The comparison should be

made with respect to unit mass of reactants, unit mass of fuel, and unit volume of fuel alone.

(b) Given that the enthalpy of vaporization of liquid methanol is $39.23\,kJ\,mol^{-1}$ and its density is $790\,kg\,m^{-3}$, also calculate the thermal energy available from the combustion of unit volume of liquid methanol. (*Note*: Some important conclusions regarding the advantages and disadvantages of different fuels may be drawn from these simple calculations.)

2. Use the simple formulae on p. 12 to estimate mean molar heat capacities at constant pressure and constant volume of carbon dioxide, water vapour and nitrogen. (Remember that the geometrical structures of CO_2 and H_2O are different). Compare the values obtained with those given by the data in Table 1.2 for temperatures of 298 and 1500 K.
3. Use the mean molar heat capacities obtained above to estimate adiabatic reaction temperatures at constant pressure and at constant volume for stoichiometric mixtures of propane in air and propane in oxygen. Compare the results with the accurate values in Tables 1.3 and 1.4.
4. Nitric oxide, NO, is an important combustion-generated pollutant. Use the relation

$$\Delta G_T^{\ominus} = -RT \ln K_p$$

to obtain an expression for the equilibrium concentration of NO in air as a function of temperature, assuming that NO, N_2 and O_2 have identical entropies at all temperatures. (How good is this assumption?). Use this expression to estimate the mole fraction of NO in air at 300, 1000, 2000 and 5000 K. What implications do these results have for the operation of internal combustion engines?

5. The following reactions all participate in the combustion of hydrogen:

	A	E
$H_2 + O_2 \rightarrow 2OH$	$2.5 \times 10^9\,dm^3\,mol^{-1}\,s^{-1}$	$163\,kJ\,mol^{-1}$
$H + O_2 \rightarrow OH + O$	$2.2 \times 10^{11}\,dm^3\,mol^{-1}\,s^{-1}$	$70.3\,kJ\,mol^{-1}$
$O + H_2 \rightarrow OH + H$	$1.7 \times 10^{10}\,dm^3\,mol^{-1}\,s^{-1}$	$39.5\,kJ\,mol^{-1}$
$OH + H_2 \rightarrow H_2O + H$	$2.2 \times 10^{10}\,dm^3\,mol^{-1}\,s^{-1}$	$21.5\,kJ\,mol^{-1}$
$H + O_2 + M \rightarrow HO_2 + M$	$1.8 \times 10^9\,dm^6\,mol^{-2}\,s^{-1}$	0

Evaluate the rate of each reaction for a stoichiometric mixture of hydrogen and oxygen at 3000 K and atmospheric pressure. Assume the active species are each present in concentrations equal to 1/100th of the total reactants. The first and last of these elementary steps have relatively small rates. Explain why they may still play an important part in the overall reaction. The other three steps form a chain cycle. Which of them would you expect to dominate the combustion behaviour of the hydrogen–oxygen system?

6. Assume that ethane pyrolyses by the following mechanism:

	A	E
$C_2H_6 \rightarrow 2CH_3$	$10^{16.9}\,s^{-1}$	$374\,kJ\,mol^{-1}$
$CH_3 + C_2H_6 \rightarrow CH_4 + C_2H_5$	$10^{9.6}\,dm^3\,mol^{-1}\,s^{-1}$	$73.2\,kJ\,mol^{-1}$
$C_2H_5 \rightarrow H + C_2H_4$	$10^{14.4}\,s^{-1}$	$171\,kJ\,mol^{-1}$
$H + C_2H_6 \rightarrow C_2H_5 + H_2$	$10^{11.1}\,dm^3\,mol^{-1}\,s^{-1}$	$39.2\,kJ\,mol^{-1}$
$2C_2H_5 \rightarrow C_4H_{10}$	$10^{8.5}\,dm^3\,mol^{-1}\,s^{-1}$	0

On this basis, show that the reaction will follow a half-order law and calculate the global Arrhenius parameters. What will be the major and minor products of this reaction? For an ethane pressure of 100 torr and a temperature of 800 K, calculate the stationary-state concentrations of the radical species, the rate of removal of ethane and the rate of formation of butane.

7. Use the expressions

$$\bar{c} = (8RT/\pi M)^{1/2} \quad \text{and} \quad l = 1/\sqrt{2}\pi n\sigma^2$$

where n is the number of molecules per unit volume, σ is the molecular diameter, and M is the molar mass, to estimate the dependence of the diffusion coefficient and the thermal conductivity on the pressure, temperature and molar mass of an ideal gas.

8. Estimate $\bar{c}$, l, D and λ for air at room temperature and atmospheric pressure, assuming $\sigma = 0.3$ nm. Hence calculate the Lewis number for air and compare it with the value obtained by using the kinetic theory expressions for a and D.

2 Explosions in closed vessels

In this chapter, the theory of explosions in closed vessels is developed in greater detail. The notion of a closed vessel is meant to imply that the whole volume is simultaneously involved in the phenomenon and we can disregard propagating combustion waves. The analysis is applicable to all phases, although we shall see later that branching-chain explosions are most often found in gaseous systems. In liquids and solids, diffusion coefficients are several powers of ten lower than in gases, while thermal conductivities are typically higher by an order of magnitude. Indeed, diffusion of active species is virtually impossible in solids until melting occurs. Consequently, explosions in condensed phases generally have their origin in thermal effects.

Thermal explosions and branching-chain explosions are usually treated quite independently, it being assumed that in the former, the chemical reaction may be characterized by a simple rate law while, in the latter, no heat is liberated by chemical reaction. Such a distinction is grossly oversimplified; nevertheless, in many cases it provides a satisfactory description of the limit conditions. However, occasionally it can lead to incorrect conclusions and it then becomes necessary to combine thermal and chain factors in a *unified theory* [15–18].

2.1 THERMAL EXPLOSIONS

2.1.1 Theory of thermal explosions

At the end of the last century it was recognized that the autocatalytic action required for an explosion can arise from the self-heating produced by an exothermic reaction because rates of chemical reactions increase dramatically with rising temperature. The criterion for occurrence of an explosion must therefore be related to the net rate of heat loss or gain in a volume element of the reacting system. If the rate of heat loss due to conduction, convection and radiation remains equal to the rate of heat generation by reaction, then a stable temperature distribution will be established, while if the rate of heat loss falls below that of heat generation an explosive situation will follow.

Semenov [19] developed an elementary model for thermal explosions which serves to demonstrate the principal features of the phenomenon. In this model, the temperature of the reacting system, T, is assumed to be uniform over the whole volume but to differ from that of the walls of the container which are at a temperature T_w (Fig. 2.1). So long as the temperature difference is not too large the rate of heat loss may be written as

$$\dot{q}_- = hS(T - T_w) \tag{2.1}$$

where h is a heat transfer coefficient and S is the surface area of the

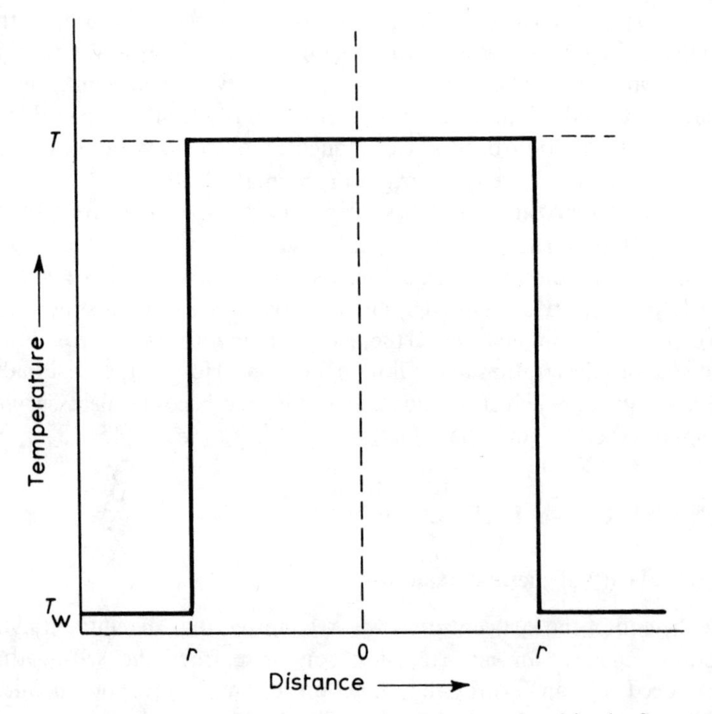

Fig. 2.1 Temperature profile in the reaction vessel assumed in the Semenov theory of thermal explosions [19].

vessel. The rate of heat production is given by

$$\dot{q}_+ = rVQ \tag{2.2}$$

where r is the rate of reaction in moles per unit volume per unit time, V is the volume of the reacting system and Q is the exothermicity of the reaction. For a process taking place at constant volume, $Q = -\Delta U$, while for a constant pressure process, $Q = -\Delta H$. If it is assumed that the rate of reaction displays the normal Arrhenius temperature dependence then

$$\begin{aligned}\dot{q}_+ &= kf(c)VQ \\ &= A\exp(-E/RT)f(c)VQ\end{aligned} \tag{2.3}$$

where $f(c)$ is the appropriate function of reactant concentration. For a given system, plots of these heat flow terms against temperature are shown in Fig. 2.2. The rate of heat loss is a straight line passing through T_w and the rates of heat production form a family of curves related to the concentrations of reactants.

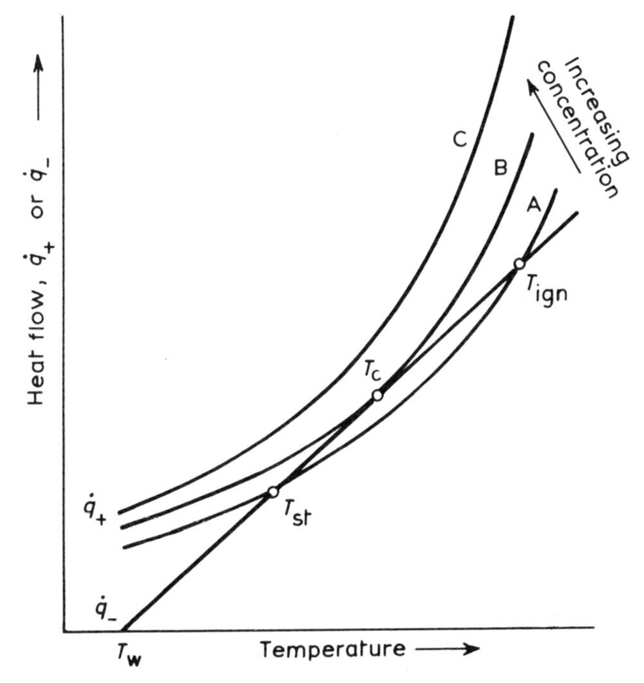

Fig. 2.2 Heat flow contributions for an exothermic reaction in a closed vessel. Curves A, B and C represent the heat production terms for different reactant concentrations.

Considering first that the reactants are admitted to the container at a temperature T_w under concentration conditions corresponding to curve A, then reaction will commence and because the heat production curve lies above the heat loss curve, the system will heat up until a temperature T_{st} is attained. At this point, the rates of heat generation and heat loss are equal and the reaction will proceed steadily without further acceleration. This intersection of the two curves corresponds to a stable situation, and if a small temperature excursion takes place in either direction the system will return to T_{st}. However, if the reactants are heated by some external source to a temperature above the second intersection, T_{ign}, the situation becomes unstable and ignition occurs. Note that a stable condition cannot be maintained at the second intersection in the way that it could at T_{st}. Thus, if the temperature rises above T_{ign}, the reaction autoaccelerates while, if it falls below T_{ign}, the system will drop back to the stable reaction condition at T_{st}. Turning now to curve C, it can be seen that here the reaction is immediately explosive since the rate of heat production always exceeds that of heat loss.

The limiting situation, which marks the boundary between the two types of behaviour, is illustrated by curve B to which the heat loss curve forms a tangent, the intersection points T_{st} and T_{ign} coinciding at T_c. The system will heat up slowly from T_w to T_c after which rapid self-heating and acceleration to explosion occur. These conditions describe the minimum requirements for *self-ignition.*

2.1.2 Thermal explosion limits

The critical conditions for thermal explosion will change if the heat loss curve is altered. This can happen in two ways. First, if T_w the temperature of the vessel walls is changed, then the curve will be displaced to the left or right. Fig. 2.3 shows that increase in the value of T_w moves more of the family of heat production curves into the explosive region. Alternatively, if the rate of heat transfer changes due to an alteration in vessel dimensions or heat transfer coefficient, the slope of the heat loss curve is modified. Fig. 2.4 shows that an effect similar to that in Fig. 2.3 is created, that is, the range of conditions under which explosion occurs is increased if the rate of heat transfer is reduced.

It is of interest to examine the limiting conditions for explosion in more detail. At the critical point, T_c, the expressions for $\dot{q}_-$ and $\dot{q}_+$ may be equated

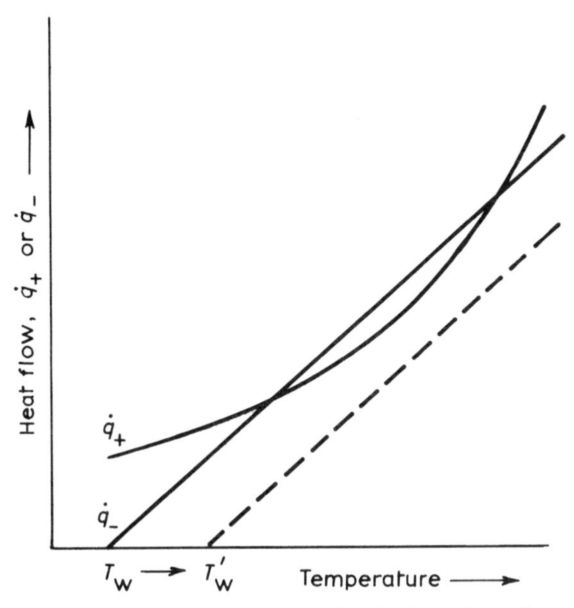

Fig. 2.3 The effect on thermal explosion limits of raising the wall temperature.

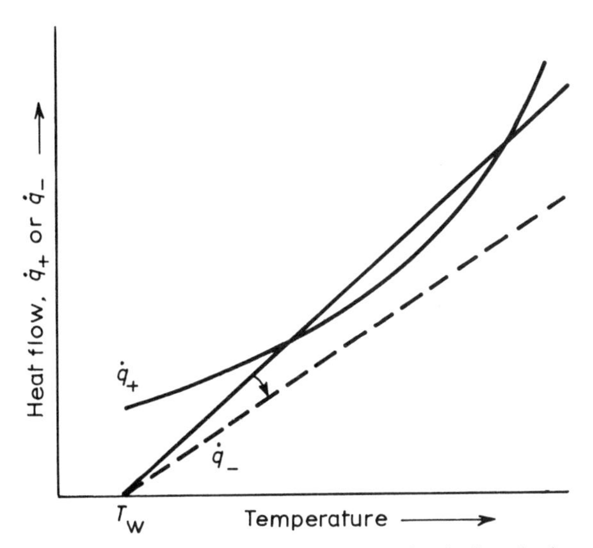

Fig. 2.4 The effect on thermal explosion limits of reducing the heat transfer coefficient.

$$hS(T_c - T_w) = A\exp(-E/RT_c)f(c)VQ \tag{2.4}$$

and, since the curves are tangential, their slopes are equal, that is,

$$\frac{d\dot{q}_-}{dT} = \frac{d\dot{q}_+}{dT} \tag{2.5}$$

Differentiating Equations 2.1 and 2.3 with respect to T gives

$$hS = A\exp(-E/RT_c)(E/RT_c^2)f(c)VQ \tag{2.6}$$

Dividing Equation 2.4 by Equation 2.6 leads to the equation

$$T_c - T_w = RT_c^2/E \tag{2.7}$$

This is a quadratic in T_c which can be solved giving

$$T_c = (E/2R)[1 \pm (1 - 4RT_w/E)^{1/2}] \tag{2.8}$$

Only the root with the negative sign is of interest, the other solution corresponding to an exceedingly high temperature and a physically unreal situation. Expanding the square root term using the binomial theorem leads to the conclusion that, to a good approximation,

$$T_c - T_w = RT_w^2/E \tag{2.9}$$

This temperature change represents the pre-explosion heating and its value will normally be quite small. For example, with $E = 200\,\text{kJ mol}^{-1}$ and $T_w = 500\,\text{K}$, the change is about 10 K. It also follows from Equation 2.9 that when explosion is about to occur, the rate of reaction has increased by a factor of e$(=2.718)$ from its initial value at T_w.

Explosion limits are usually described in terms of a concentration–temperature relationship. The concentration corresponding to the critical condition may be obtained by substituting the expression for T_c in Equation 2.4. Subsequent manipulation leads to the expression

$$f(c)_c = \frac{hSRT_w^2}{EAVQe}\exp(E/RT_w) \tag{2.10}$$

or

$$\ln[f(c)_c/T_w^2] = (E/RT_w) + \ln B \tag{2.11}$$

where $B = hSR/EAVQe$.

For a gas of partial pressure p, the corresponding concentration c is given by $c = p/RT$ and thus, for a simple gas reaction of order j, the pressure limit is defined by

$$\ln(p^j/T_w^{2+j}) = (E/RT_w) + \ln B' \tag{2.12}$$

Many explosions are found to obey a relationship of this type. However, since branching-chain explosions can be shown to fit a very similar relationship at their lower limit, it cannot be used as a criterion for determining whether a particular explosion arises because of thermal effects or radical-chain branching. More detailed theories of thermal explosions do provide relationships between vessel dimensions and vessel temperature which can be verified experimentally and used to establish the nature of the explosion.

Once the thermal explosion has commenced, the rate of change of temperature will be determined by the difference between the heat generation and heat loss terms divided by the heat capacity of the system. Since the former increases as $\exp(-E/RT)$ and the latter simply with T, the temperature increases at a rate close to exponential as explosion proceeds.

The Semenov theory provides a qualitative understanding of the nature of critical conditions for explosion and leads to expressions for the pre-explosion temperature rise and for the way that the critical concentration varies with temperature. Its main limitation is the assumption of a uniform distribution of temperature throughout the reactants. While such an approximation will be satisfactory if the reactants have a high thermal conductivity and the walls are thermally insulating, or if the reactant fluid is vigorously stirred, we must also

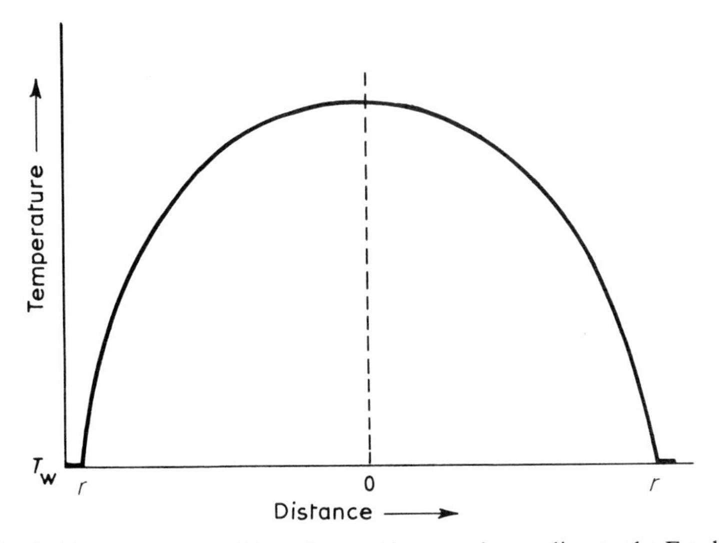

Fig. 2.5 Temperature profile in the reaction vessel according to the Frank-Kamenetskii [20–24] theory of thermal explosions.

consider the opposite extreme, that is, reactants with a relatively low thermal conductivity in a vessel with highly conducting walls. This situation is illustrated in Fig. 2.5, and has been treated by Frank-Kamenetskii [20–24]. In the stationary state, the differential heat balance may be written

$$\lambda \nabla^2 T + Q f(c) A \exp(-E/RT) = 0 \tag{2.13}$$

with the boundary conditions $T = T_w$ at the wall and, for a symmetrical vessel, $dT/dr = 0$ at the centre; ∇^2 is the Laplacian operator $(\partial^2/\partial x^2 + \partial^2/\partial y^2 + \partial^2/\partial z^2)$; and Q is the exothermicity of the reaction. Solutions to this equation exist only when a dimensionless parameter, δ, is less than a certain value. Solutions in the form of stable temperature profiles $T = f(r)$ are then possible and, under these circumstances, steady reaction takes place with a certain temperature distribution defined by $f(r)$ in the reacting medium. Above the critical value of δ, no stable temperature profile is possible and explosion occurs.

The parameter δ is actually the dimensionless heat release rate appropriate to the temperature of the surroundings and its value is given by

$$\delta = QEr^2 A \exp(-E/RT_w) f(c)/\lambda R T_w^2 \tag{2.14}$$

The theory also leads to an expression for the maximum pre-explosion temperature rise, ΔT_c, at the centre of the vessel; this differs from that predicted by the Semenov theory by a numerical factor of order unity. The temperature profile in the non-explosive case is close to, but not precisely, parabolic.

The value of δ at the explosion limit and the corresponding values of $\Delta T_c = T_c - T_w$ depend on the geometry of the system and the results for three simple cases are given in Table 2.1.

The limiting expression for explosion in the Semenov theory is given by Equation 2.10 which can be rearranged to

$$1/e = QEVA \exp(-E/RT_w) f(c)/hSRT_w^2 \tag{2.15}$$

Comparison of Equations 2.14 and 2.15 shows that the two theories have a certain formal similarity. Both predict that explosion will occur when a dimensionless group, all of whose constituent terms are in principle known, exceeds a certain number. In the Semenov theory the critical value is $1/e = 0.368$, while in the Frank-Kamenetskii theory the critical number depends on the geometry and is, for example, 3.32 for a sphere. While the two theories lead to the same dependence on T_w, E

Table 2.1 Values of δ_c and ΔT_c for various vessel geometries according to the Frank-Kamenetskii theory.

Shape of vessel	δ_c	ΔT_c	Characteristic dimension, r
Infinite slab	0.88	$1.20\,RT_w^2/E$	Half-width of slab
Infinite cylinder	2.00	$1.37\,RT_w^2/E$	Radius of cylinder
Sphere	3.32	$1.60\,RT_w^2/E$	Radius of sphere

and Q, there is a crucial difference in the scale effect; the Semenov treatment leads to a dependence on $1/r$, while the Frank-Kamenetskii expression is proportional to $1/r^2$. It also follows from Equations 2.14 and 2.15 that the two treatments become identical if

$$hS/V = \delta_c \lambda e/r^2 \tag{2.16}$$

This expression can be rearranged to

$$h = b(\lambda/r) \tag{2.17}$$

where $b = 2.39$, 2.72 and 3.01 for a slab, cylinder and sphere, respectively.

The Frank-Kamenetskii analysis retains a number of features implicit in the Semenov theory. No allowance is made for consumption of reactants or for dealing with vessels of complex geometry. These factors are taken into account by extensions to the original theory [25–29]. Heat transfer is taken to occur only by conduction whereas, in gases at least, convective heat transfer is important under some conditions. In experiments with gases in closed vessels, convection begins to be important when the Rayleigh number exceeds 600.* This means that in gaseous systems convection is negligible at pressures below about 0.1 atm, but above this pressure heat transfer is enhanced (and explosion correspondingly hindered) by convective forces [30].

Thermal explosions in solids are often generated when the surface of the material or of a thin-walled container is exposed to the atmosphere. The heat loss is then fairly well represented by a surface transfer coefficient, the major temperature gradients lying close to the walls, and the simpler Semenov treatment often holds quite well.

* The Rayleigh number is a dimensionless group defined by

$$Ra = \boldsymbol{g}\alpha r^3 \Delta T \rho^2 c_p / \lambda \eta$$

where $\boldsymbol{g}$ is the gravitational acceleration, α is the coefficient of thermal expansion, and η is the dynamic viscosity.

Whichever treatment is appropriate for any particular set of conditions (temperature, density, geometry), there is a critical dimension below which explosion cannot occur. This has an immediate consequence when considering hazards to safety and there are many examples where thermal explosion theory has been usefully applied to practical problems [31].

Another consequence of the effect of scale is concerned with initiation. In a solid, combustion is usually initiated by the formation of a local hot spot. This may be brought about in a variety of ways, often mechanical in nature; for example, frictional heating, adiabatic compression of trapped gas bubbles and shock compression can all transform the mechanical disturbance into a small zone of increased temperature. Provided the size of the region and the temperature achieved are together sufficient to sustain self-heating, the hot spot necessary for ignition will be present. A typical hot spot is likely to have a diameter of 10^{-3}–10^{-5} cm, a temperature of 500° C, and a duration of 10^{-4}–10^{-6} s.

2.2 BRANCHING-CHAIN EXPLOSIONS

2.2.1 Chain reactions

In branching-chain explosions, autocatalysis occurs because of the formation of highly reactive chemical entities, or active centres, which can themselves react further to produce more such species. Active centres are usually free atoms or radicals, although in certain systems molecules which are unstable or vibrationally or electronically excited may play a similar role. The reactions leading to the initial formation of active centres are highly endothermic and therefore slow. However, the centres react rapidly with stable molecules to give more similar species. The complete sequence is termed a *chain reaction.*

In a chain reaction, four main types of process can be identified:

(a) *Initiation.* In this process, atoms or radicals are produced by the dissociation of either a reactant molecule or some substance (an *initiator*) added specifically to promote initiation. We are concerned in this section primarily with *spontaneously explosive* systems and therefore we need to consider only the production of atoms or radicals by thermal dissociation. Branching-chain explosions can occur outside the limits for spontaneous explosion when initiation is brought about by some external means, for example, a high-energy light source, an electric spark etc. A unimolecular dissociation reaction will have an activation energy equal to, or greater than, the bond dissociation energy, say 200–500 kJ mol^{-1}, and the rate will therefore be slow. Bimolecular initiation reactions tend to have lower activation energies and lower pre-exponential factors and hence are only rarely significantly faster. Initiation may take place homogeneously or heterogeneously, particularly since the walls of the container may catalyse the reaction. Formally we may write

$$A \rightarrow 2X \qquad \text{rate, } I_0 \tag{2.19}$$

where A denotes an initial reactant, X an active species and I_0 is an appropriate function of the rate constant and concentration which gives the number of active species formed in unit volume per unit time.

(b) *Propagation.* The propagation reaction is important because it governs the rate at which the chain continues. The requirement is for an active centre X to react with a molecule producing a second active centre Y, for example,

$$X + A' \rightarrow Y + P \qquad \text{rate, } f_p[X] \tag{2.20}$$

where A′ is a molecule and P is a stable product. Throughout this analysis, the quantity f contains all the rate constants and concentration terms which make up the reaction rate, apart from the concentration of active centres; thus, for processes such as Reaction 2.20 it takes the form of a first-order coefficient with dimension of [time]$^{-1}$. It is unlikely that X and Y will be identical so the chain normally propagates by a 'shuttle' between two or more active centres

$$Y + A'' \rightarrow X + P \tag{2.21}$$

For most propagation reactions of importance in combustion, activation energies lie between zero and 45 kJ mol^{-1}.

In the later stages of a reaction as the product concentration increases markedly a propagation reaction may appear to 'reverse' thus regenerating reactant and causing inhibition

$$Y + P \rightarrow X + A' \tag{2.22}$$

For the purpose of the present analysis, this can still be regarded as belonging to the category of propagation reactions.

(c) *Branching.* The branching reaction is a special type of propagation reaction in which two or more active centres are formed

$$X + A' \rightarrow 2Y + \cdots \tag{2.23}$$

This is the reaction which is responsible for the explosion. Since it need not necessarily occur very rapidly, its activation energy may be appreciably higher than that of the ordinary propagation reaction with which it competes. The formal reaction above corresponds to *normal branching* but other types are possible. Highly energetic species may transfer energy to a stable molecule which is then able to dissociate: this is termed *energy branching* and is believed to occur in the oxidation of carbon monoxide via excited CO_2 molecules (Chapter 5). The active centres produced in the branching reaction are not necessarily the same as those involved in the ordinary propagation steps, but nevertheless they are capable of containing the chain reaction. Sometimes branching involves decomposition of a stable product to give active centres

$$P \rightarrow X + \cdots \tag{2.24}$$

This is a relatively slow process and it is therefore termed *degenerate* (or *delayed*) branching (Chapter 6).* Finally, two active centres may react together giving quadratic branching, for which the kinetic term becomes $k_b'[X]^2$ but evidence for quadratic branching in most systems is very limited.

(d) *Termination.* A combination of initiation and propagation steps would cause the overall reaction rate to accelerate without limit, even in the absence of chain branching, were it not for competition for the active centres by reactions which terminate the chains. Gas-phase termination occurs either by recombination of two radicals to give a stable molecule, or by reaction of a radical with a molecule to give either a molecular species or a radical of lower reactivity which is unable to propagate the chain. Since both these processes will be exothermic, a third body is usually required to take up the energy released and prevent redissociation

$$X + X + M \rightarrow X_2 + M \quad \text{rate}, f_g[X]^2 \tag{2.25}$$

$$X + R + M \rightarrow XR + M \quad \text{rate}, f_g[X] \tag{2.26}$$

Removal of radicals at the wall can take place by a variety of processes, the details of which are usually unimportant as the rate-controlling step is normally diffusion through the gas

$$X \rightsquigarrow \tfrac{1}{2}X_2 \quad \text{rate}, f_w[X] \tag{2.27}$$

By analogy with chain branching, Reactions 2.26 and 2.27 are referred to as linear termination and Reaction 2.25 as quadratic termination. Strictly speaking, f_w is rather different from the previous rate factors

* Some authors use the term *secondary initiation* to describe this process.

and should take the form $D\nabla^2[X]$ where D is the diffusion coefficient and ∇^2 is the Laplacian operator introduced previously. However, it has been shown that identical results are obtained using the term f_w, while noting that it contains quantities to allow for the vessel geometry.

The kinetics of branching-chain reactions can be handled mathematically in two ways. To begin with, we shall attempt a direct solution of the appropriate differential equation. To achieve this, further simplification is required: we therefore neglect all quadratic terms, denote all active centres by X, and combine the propagation, branching and termination steps in a single equation

$$X + A \rightarrow \chi X + P \quad \text{rate, } \phi[X] \tag{2.28}$$

χ, the *multiplication factor* takes the following values for the three separate processes

$\chi > 1$, chain branching
$\chi = 1$, chain propagation
$\chi = 0$, chain termination

ϕ, the *net branching factor*, contains the separate quantities f_b, f_g and f_w: it is equal to $(\chi - 1)/\tau$ where τ is the mean lifetime of the active centres. We can now write the differential equation which governs the time dependence of the concentration of active centres

$$\frac{d[X]}{dt} = I_0 + \phi[X] \tag{2.29}$$

Assuming that $[X] = 0$ at $t = 0$ and that ϕ is independent of time (and therefore of any changes in temperature and concentration which may follow), this equation may be integrated to give

$$[X] = \frac{I_0}{\phi}(\exp \phi t - 1) \tag{2.30}$$

This function is plotted in Fig. 2.6 which shows that, for positive values of ϕ, the active centres grow exponentially, thus leading to explosion. For negative values of ϕ, the active centres reach a steady-state concentration $-I_0/\phi$ corresponding to a constant reaction rate and $d[X]/dt = 0$. The boundary between explosion and slow reaction is given by $\phi = 0$, and in this case integration of Equation 2.29 leads to the result

$$[X] = I_0 t \tag{2.31}$$

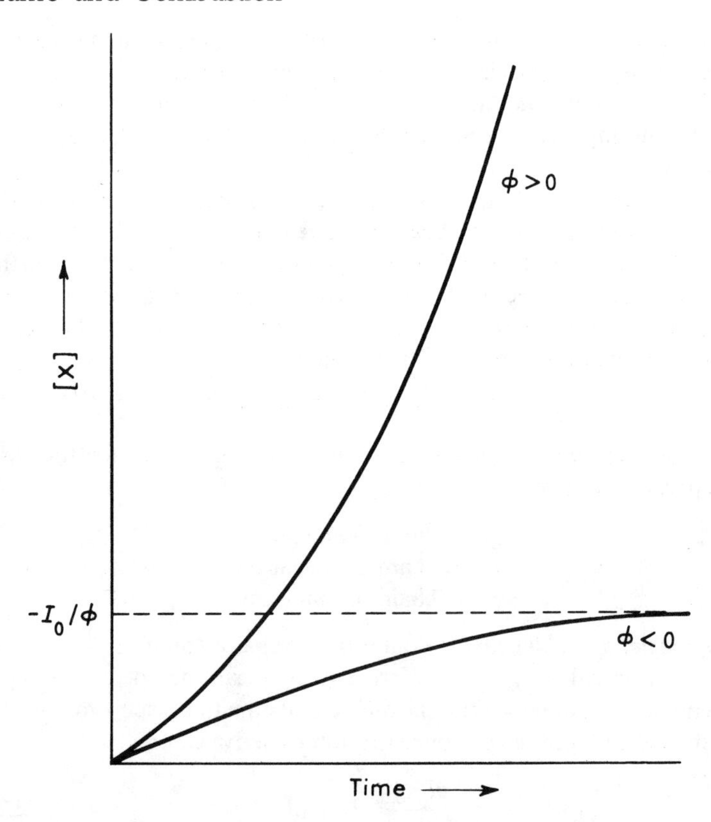

Fig. 2.6 Growth of chain-carrier concentration, [X], for different values of the net-branching coefficient, ϕ.

Inspection of Equations 2.30 and 2.31 shows that altering I_0 does not affect the explosion boundary and for this reason, it has often proved difficult to establish the initiating step in isothermal branched-chain reactions. However, small changes in any of the quantities which make up ϕ can change the sign of ϕ and so convert a slow reaction into an explosion or vice versa.

The above analysis, while qualitatively sound, is unsatisfactory because of the number of approximations made. In order to investigate the limiting behaviour it is preferable to set up equations which determine the steady-state behaviour, with $d[X]/dt = 0$, and then investigate the circumstances under which the steady state can no longer be maintained. If we rewrite Equation 2.29 in the form

$$\frac{d[X]}{dt} = I_0 + (\chi - 1)f_b[X] - (f_g + f_w)[X] = 0 \tag{2.32}$$

it follows that

$$[X] = \frac{I_0}{(f_g + f_w) - (\chi - 1)f_b} \tag{2.33}$$

and the overall rate in the steady state is

$$f_p[X] = \frac{f_p I_0}{(f_g + f_w) - (\chi - 1)f_b} \tag{2.34}$$

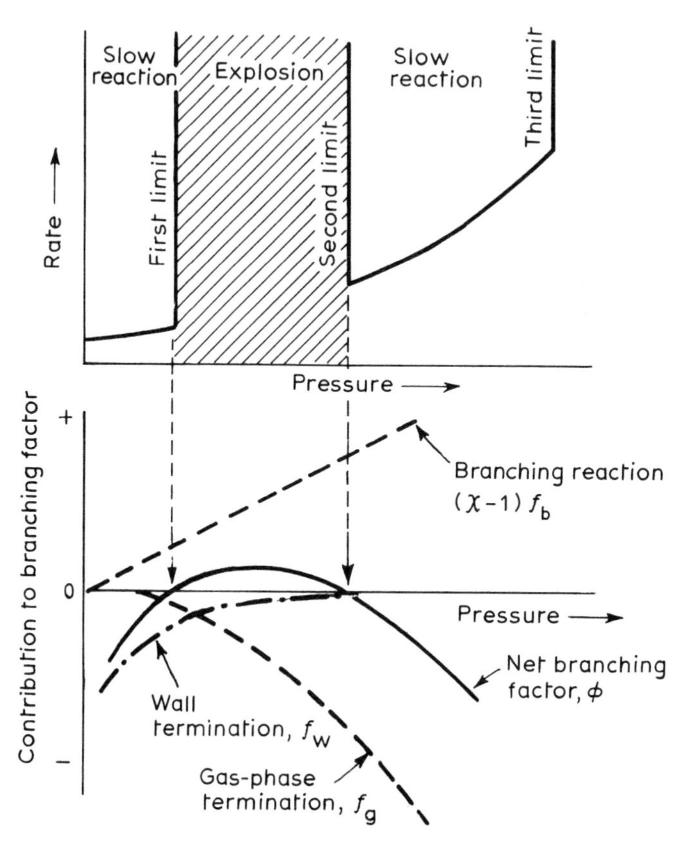

Fig. 2.7 Explosion limits in a branching-chain reaction. The lower figure shows the effect of pressure on the various contributions to the net branching factor.

From this it can be seen that if $\chi < 1$, a steady-state situation is always possible but if $\chi > 1$ the steady-state analysis breaks down when the denominator reaches zero. Under these circumstances a steady state becomes impossible and explosion occurs when

$$(\chi - 1)f_b \geqslant f_w + f_g \tag{2.35}$$

(and not simply when $\chi > 1$).

The functions f_w, f_g and f_b are likely to contain very complex pressure, temperature and composition dependences, but in general we can see that

(a) f_w will vary roughly as 1/(pressure), because of the dependence of diffusion to the walls,
(b) f_b will depend on $(\text{pressure})^1$, corresponding to reactant concentration, and
(c) f_g will involve $(\text{pressure})^2$, because of the third-body effect.

Thus we can predict the existence of two separate explosion limits, one of which occurs at low pressures and corresponds to f_g negligible and $(\chi - 1)f_b = f_w$ and the other at higher pressures where f_w may be neglected and $(\chi - 1)f_b = f_g$. The contributions are illustrated schematically in Fig. 2.7.

In conclusion, it should be emphasized that this treatment strictly refers to *chain-isothermal* reactions. Most chain reactions are exothermic overall and autoacceleration may occur because of self-heating as well as chain branching. If the thermal effect influences the position of the limits, the reaction is said to be *chain-thermal.*

2.2.2 Branching-chain explosion limits

A special feature of explosive reactions is that they are limited to well-defined ranges of temperature and composition, outside which the rate of reaction may be slow or even negligible. This effect has already been illustrated in the upper part of Fig. 2.7 where the rate of reaction is shown as a function of pressure at a fixed temperature and composition. An alternative representation (Fig. 2.8) shows the limiting behaviour of the hydrogen–oxygen reaction in terms of both temperature and pressure: the upper part of Fig. 2.7 may be regarded as a vertical 'slice' through such a plot. The same limits appear to be characteristic of all gas-phase branching-chain explosions; that is there is a *first* (or *lower*) limit of pressure above which explosion occurs and a *second* (or *upper*) limit below which explosion takes place. In many

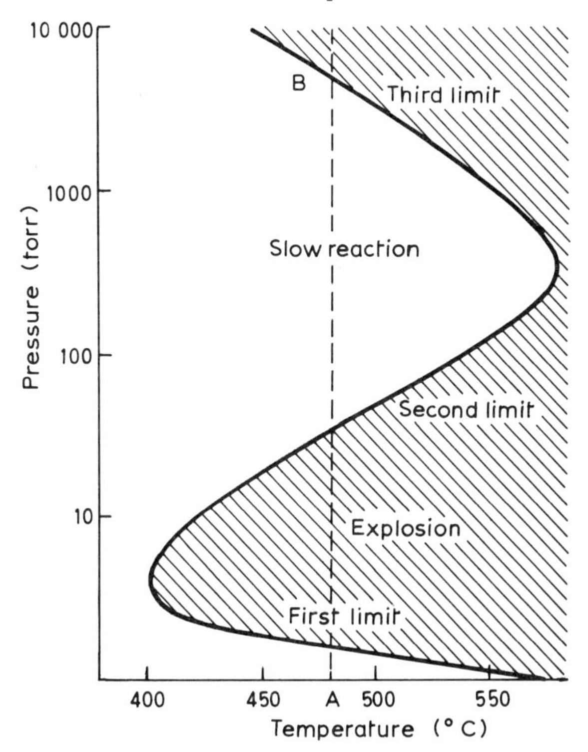

Fig. 2.8 The explosion boundary for the hydrogen–oxygen reaction as a function of temperature and pressure (after Lewis and Von Elbe [32]).

cases there is also a *third* limit above which explosion is observed. Identification of this limit is sometimes difficult because the 'slow' reaction is fast in this region and the transition to explosion may not be detected easily.

These explosion limits are important not only in themselves but because they throw considerable light on the processes which are taking place. The example of the hydrogen–oxygen reaction will be used as a basis for discussion because the mechanism is now well established [33]. However, similar behaviour is observed for a wide range of gaseous oxidations including those of carbon monoxide, carbon disulphide, hydrogen sulphide, phosphorus, phosphine and ammonia.

(a) *The first, or lower, limit*

The position of the lower limit is very susceptible to the nature of the surface coating of the container and moves to lower pressures if the size of the vessel is increased or inert gases are added. The conclusion to be drawn from this evidence is that a homogeneous chain-branching reaction competes with termination processes taking place on the vessel walls. Any change in the system which retards diffusion of chain carriers to the walls will therefore have the effect of lowering the explosion limit, and any change in the surface which alters its efficiency in removing chain carriers will also influence the limit.

The situation in the hydrogen–oxygen system can be analysed without invoking a detailed reaction mechanism [34]. It is only necessary to assume that the active centres may be classified into two types characterized by the identity of the molecule with which they react. For example, X_A reacts with hydrogen to produce X_B, and X_B reacts with oxygen to give X_A. The number of collisions Z_{A,H_2} which species X_A makes with hydrogen then depends on the ratio of the partial pressure of hydrogen p_{H_2} to the total pressure p_{total}, i.e.

$$\frac{Z_{A,H_2}}{\sum Z_A} = \frac{p_{H_2}}{p_{total}} \tag{2.36}$$

where $\sum Z_A$ represents the sum of collisions made by X_A with all species present; similarly

$$\frac{Z_{B,O_2}}{\sum Z_B} = \frac{p_{O_2}}{p_{total}} \tag{2.37}$$

But, since X_B only forms when X_A reacts and vice versa,

$$Z_{B,O_2} = Z_{A,H_2} \tag{2.38}$$

and so

$$\sum Z_A + \sum Z_B = \left\{ \frac{p_{total}}{p_{H_2}} + \frac{p_{total}}{p_{O_2}} \right\} Z_{A,H_2}$$

or

$$Z_{A,H_2} = \frac{p_{H_2} p_{O_2} Z}{p_{total}(p_{H_2} + p_{O_2})} \tag{2.39}$$

where Z is the total number of collisions suffered by the active species.

The number of collisions experienced by a molecule in diffusing a distance d is shown by the Smoluchowski equation to be proportional to d^2/l^2, where l, the mean free path, is inversely proportional to pressure. If it is assumed that active centres are destroyed at the walls and d is taken as the vessel diameter, then

$$Z_{A,H_2} \propto \frac{p_{H_2} p_{O_2} p_{total} d^2}{(p_{H_2} + p_{O_2})} \tag{2.40}$$

If the probability of a branching collision is denoted by ω, explosion may be expected if $\omega Z_{A,H_2} > 1$, that is, if more centres are formed than are lost at the walls. Thus, at the first limit, the condition

$$p_{H_2}p_{O_2}\frac{p_{total}}{(p_{H_2}+p_{O_2})}d^2 = \text{constant} \tag{2.41}$$

should be obeyed. If an inert gas is also present this may be rewritten

$$p_{H_2}p_{O_2}\left(1+\frac{p_{inert}}{p_{H_2}+p_{O_2}}\right)d^2 = \text{constant} \tag{2.42}$$

This relation has been verified experimentally for variation of both gas composition and vessel dimensions.

The effect of temperature has not been considered explicitly in this treatment. In general, diffusion coefficients vary approximately as $T^{1.5}$ while branching-chain reactions, which dictate the value of ω, have appreciable activation energies, so that the limit moves to lower pressures as the temperature is raised.

The analysis has been based on the assumption that surface termination is very efficient. The concentration profile of active centres through the vessel then closely resembles the temperature distribution resulting from the Frank-Kamenetskii treatment of thermal explosions illustrated in Fig. 2.5. One hundred per cent efficiency of removal is not necessary as, to a good approximation, a lower efficiency merely introduces a numerical factor into the relationship. However, if surface removal is very inefficient, the concentration gradients become negligible and diffusion is no longer rate determining. Instead, the importance of wall reactions relative to gas-phase reactions will depend on the surface:volume ratio and the explosion limit varies with $1/d$ rather than $1/d^2$ [35]. This situation compares directly with the Semenov model of thermal explosions and the active centre profile resembles that of the temperature in Fig. 2.1.

The question of intermediate efficiencies has been examined in detail by Baldwin [36] who has shown that a surface efficiency of 10^{-3} may give termination independent of diffusion at first limit pressures below 10 torr, but with almost complete diffusion control during slow reaction at pressures of several hundred torr.

(b) *The second, or upper, limit*

While the first limit is typically at pressures of a few torr, the second limit is observed at considerably higher pressures. At the second limit, increase in pressure inhibits the explosive reaction. This means that some process competes with chain branching and shows a higher order dependence on pressure. The dimensions of the vessel and the nature of its surface have only a minor effect on the reaction controlling chain branching which operates in the gas phase.

In the hydrogen–oxygen reaction the major chain-branching step is

$$H + O_2 \rightarrow OH + O \tag{2.43}$$

(see Chapter 5). However, if a third body is available to remove the excess energy then a different process may occur

$$H + O_2 + M \rightarrow HO_2 + M \tag{2.44}$$

The HO_2 radical is less reactive than the other active intermediates involved in the chain – H, O and OH – and hence has a greater probability of travelling to the wall and being destroyed there. The efficiency of Reaction 2.44 in terminating the chain will depend to some extent on those factors which control the destruction of HO_2 at the walls, and this explains the small effect of the nature of the surface and the vessel geometry.

The competing reactions may be written quite generally [37] as

$$X + A \xrightarrow{k_1} \chi X + \cdots \quad \text{chain branching} \tag{2.45}$$

$$X + A + M \xrightarrow{k_2} XA + M \quad \text{chain termination} \tag{2.46}$$

Then, at the limit, the rate of production of radicals by branching is just balanced by their removal, and so

$$\chi k_1[X][A] = k_1[X][A] + k_2[X][A][M]$$

or

$$\begin{aligned}(\chi - 1)k_1[X][A] &= k_2[X][A][M]\\ &= k_2[X][A]\{[M_A] + \beta_B[M_B] + \beta_C[M_C] + \cdots\}\end{aligned} \tag{2.47}$$

where β_B, β_C... are the collision efficiencies of M_B, M_C etc. as third bodies relative to M_A. Writing concentrations in terms of partial pressures leads to the result, for a fixed temperature,

$$p_{M_A} + \beta_B p_{M_B} + \beta_C p_{M_C} + \cdots = \text{constant} \tag{2.48}$$

at the second limit. This expression has been verified experimentally for hydrogen, oxygen, nitrogen and some other inert gases. Moreover, the relative efficiencies β_B, β_C... are found to be close to the ratios of the collision numbers calculated from momentum transfer cross-sections.

The constant in Equation 2.48 includes the ratio k_1/k_2 and since E_2 is close to zero (or negative), while E_1 is positive, the limit moves to higher pressures as the temperature is raised.

(c) *The third limit*

There is insufficient evidence to show whether a third explosion limit is a characteristic of all branching-chain explosions and even less, in most cases, to establish the mechanism responsible. Certainly the rate of reaction, (and therefore the rate of heat release), is usually quite high immediately below the third limit, and the characteristics observed tend towards those associated with thermal explosions. For example, the influence of additives corresponds more with their heat transfer properties than with their effect on diffusion.

In the hydrogen–oxygen system, an isothermal third limit is, in principle, possible. Such a limit would arise in the following way. Above the second limit the HO_2 radicals formed in Reaction 2.44 diffuse to the wall where they are deactivated. At still higher pressures, this diffusion process is hindered and if the HO_2 radicals can react with hydrogen

$$HO_2 + H_2 \rightarrow H_2O_2 + H \tag{2.49}$$

chain carriers will be regenerated and the reaction will again become explosively fast. The third limit has been interpreted in this way by some workers [38], but under most conditions it is not possible to neglect self-heating completely.

2.3 THE CHAIN-THERMAL, OR UNIFIED, THEORY

The so-called *unified theory* recognizes that when reaction rates become appreciable, some self-heating is inevitable and the distinction between thermal and branching-chain mechanisms becomes meaningless. In other words, the mass and energy conservation equations applicable to the reacting system are coupled and must be solved simultaneously [15–18].

The essentials of the unified theory can be appreciated by considering a very simple system with first-order branching and termination reactions

$$X + \cdots \rightarrow 2X + \cdots \qquad k_b, \Delta H_b \tag{2.50}$$

$$X + \cdots \rightarrow \text{Inert products} \qquad k_t, \Delta H_t \tag{2.51}$$

where X is the chain carrier.

Neglecting the initiation process, the concentration of chain carriers is given by

$$\frac{d[X]}{dt} = (k_b - k_t)[X] \tag{2.52}$$

The energy conservation equation is

$$\frac{dT}{dt} = (1/\bar{C})\{k_b[X]\Delta H_f + k_t[X]\Delta H_t\} + \dot{q}_- \tag{2.53}$$

where $\bar{C}$ is the average molar heat capacity of the system and $\dot{q}_-$ is the heat loss term; for small temperature differences between the system and surroundings, this will be given, as in Equation 2.1, by

$$\dot{q}_- = hS(T - T_w) \tag{2.1}$$

The simultaneous solution of Equations 2.52 and 2.53 to give [X] and T as explicit functions of t is extremely difficult and a different approach is used. The independent variable t is eliminated by division of Equation 2.53 by Equation 2.52 giving

$$\frac{dT}{d[X]} = \frac{(1/\bar{C})\{k_b[X]\Delta H_b + k_t[X]\Delta H_t\} + \dot{q}_-}{(k_b - k_t)[X]} \tag{2.54}$$

and the behaviour of the system in the T–[X] plane is examined. In this plane, $dT/d[X]$ will be defined at every point except where both $d[X]/dt$ and dT/dt are simultaneously zero. These points are called singular points, or *singularities*, and they represent equilibrium states to which the system will tend. The behaviour of solutions of Equation 2.54 near the singularities can be investigated by the standard mathematical methods of non-linear mechanics [39, 40]. Under certain circumstances the system approaches equilibrium monotonically while under others the system may oscillate in various ways [41, 42].

Using this approach it is possible to give an account of all the ignition phenomena already described as well as some additional features which are peculiar to hydrocarbon oxidation; these include periodic cool flames, negative temperature coefficient of the rate and multistage ignition. Other points also emerge from the theory, one of the more surprising being that, when the effect of self-heating is small, temperature excesses may pass through their maxima and be decreasing immediately prior to explosion; when explosion occurs it is, of course, accompanied by the rapid rise in temperature characteristic of ignition. Falling temperatures immediately before an explosion have now been observed in at least two systems [43, 44].

The unified theory includes both the thermal and isothermal branching-chain theories as limiting cases; thus if $k_b = k_t$, only thermal factors are involved, while if $\Delta H_b = \Delta H_t = 0$, the system is isothermal. Reactant temperature and composition are assumed to be uniform and reactant consumption is ignored. Experiments in closed vessels represent a very poor approximation to these conditions and much of the experimental work to test the unified theory has been carried out in continuously stirred flow reactors. In these, the complex behaviour of many hydrocarbon oxidations can be studied under conditions much closer to those assumed in the theoretical development [45].

2.4 SUGGESTIONS FOR FURTHER READING

Baldwin, R.R. and Walker, R.W. (1972) Branching-chain reactions: the hydrogen–oxygen reaction, in *Essays in Chemistry* (ed. J.N. Bradley, R.D. Gillard and R.F. Hudson) vol. 3, Academic press, London.

Dainton, F.S. (1966) *Chain Reactions*, 2nd edn, Methuen, London.

Gray, P. and Lee, P.R. (1967) Thermal explosion theory, in *Oxidation and Combustion Reviews* (ed. C.F.H. Tipper) Vol. 2, Elsevier, Amsterdam.

Hinshelwood, C.N. (1940) *The Kinetics of Chemical Change*, Oxford University Press, Oxford.

Mulcahy, M.F.R. (1973) *Gas Kinetics*, Nelson, London.

Semenov, N.N. (1935) *Chemical Kinetics and Chain Reactions*, Oxford University Press, Oxford.

2.5 PROBLEMS

1. The Texas City disaster occurred because bags filled with hot, damp ammonium nitrate fertilizer were packed tightly together in a ship's hold. Given a rate constant for the decomposition of ammonium nitrate of $6 \times 10^{13} \exp(-170\,\mathrm{kJ\,mol^{-1}}/RT)\,\mathrm{s^{-1}}$, an enthalpy of formation, $\Delta_f H = -378\,\mathrm{kJ\,mol^{-1}}$, a density of $1750\,\mathrm{kg\,m^{-3}}$, and a thermal conductivity of $0.126\,\mathrm{W\,m^{-1}\,K^{-1}}$, use the Frank-Kamenetskii theory to estimate the critical radius for self-ignition at 25 and 100° C.
2. Assuming that the ammonium nitrate in the previous question was stored at 100° C, estimate the increase in temperature prior to ignition.
3. The decomposition of diethylperoxide has been studied by Fine *et al.* [46]. In a spherical reactor at 183.6° C the explosion limit of pure diethylperoxide vapour was 6.1 torr, and in experiments in which slow reaction, but not explosion, occurred the temperature rise at the centre of the vessel never exceeded 20° C. Are these results consistent with explosions occurring by a purely thermal mechanism?

   Data
   Rate constant, $k = 10^{14.2} \exp(-143\,\mathrm{kJ\,mol^{-1}}/RT)\,\mathrm{s^{-1}}$
   Thermal conductivity, $\lambda = 0.027\,\mathrm{W\,m^{-1}\,K^{-1}}$

Enthalpy change in reaction, $\Delta H = -197\,\mathrm{kJ\,mol^{-1}}$
Radius of reactor, $r = 60.6\,\mathrm{mm}$

Consult the original paper for an account of more exhaustive tests of the theory.

4. On the basis of the Semenov theory of thermal explosions, show that in the period from commencement of reaction up to the on-set of explosion, the rate of reaction has increased by a factor of e(= 2.718).
5. Under a particular set of conditions, the rate of a combustion reaction increased as shown below:

Time (s)	2	4	6	8	10.
Rate (*torr* s^{-1})	2.7	7.4	19.1	55.6	148.4

Assuming that the reactant temperature did not change and that $t > 1/\phi$, calculate the net branching factor.

3 Flames and combustion waves

In this chapter we shall discuss the commonest example of a combustion process – *the flame*. Here again the phenomenon depends on the interplay of chemical and physical processes, the present chapter dealing primarily with the latter. A flame is caused by a self-propagating exothermic reaction which is usually associated with a luminous reaction zone. It will propagate through a stationary gas at a characteristic velocity termed the *burning velocity*, or it may be held at a standstill by forcing the reactant gas to move towards the flame front at the same speed. In general, the second type of flame is not particularly stable but may be made so by use of a suitable *burner*.

The distinction between pre-mixed flames and diffusion flames has already been explained. The major part of the discussion will be concentrated on the former, as they are better understood.

3.1 PRE-MIXED FLAMES

3.1.1 Properties of pre-mixed flames

The properties of the flame will first be considered in an idealized situation where the effect of the container may be neglected, for example, in a long tube of uniform cross-section. The gas flow is taken to be uniform across the diameter so the flame front is planar and perpendicular to the flow. The gas flow rate is adjusted to make the flame stationary and is therefore equal in magnitude but opposite in sign to the burning velocity.

If we consider two planes, one ahead of and one behind the flame (Fig. 3.1), then the following conservation relationships can be applied

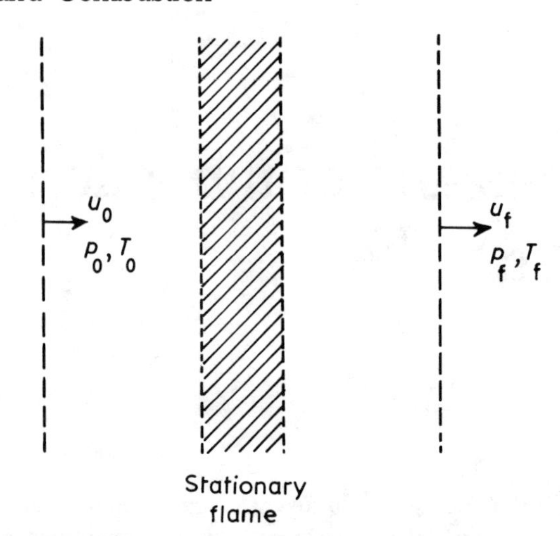

Fig. 3.1 Notation used in the description of a stationary flame.

Conservation of mass:

$$\rho_0 u_0 = \rho_f u_f = \dot{m} \tag{3.1}$$

where $\dot{m}$ is the mass flow rate per unit area through the transition, and u is the velocity of the gas stream.
Conservation of momentum:

$$p_0 + \rho_0 u_0^2 = p_f + \rho_f u_f^2 \tag{3.2}$$

Rearrangement of these relationships gives

$$\frac{(p_f - p_0)}{(1/\rho_f) - (1/\rho_0)} = -\dot{m}^2 \tag{3.3}$$

$$\frac{(p_f - p_0)}{(u_f - u_0)} = -\dot{m} \tag{3.4}$$

These expressions tell us, first of all, that the pressure and the density must change in the same direction. There are thus two types of process possible: *detonations*, in which both pressure and density increase across the transition, and *deflagrations* (i.e. combustion waves and flames) in which pressure and density decrease across the transition. Detonations are discussed in more detail in Chapter 4; they are supersonic compression waves in which chemical reaction is brought about by compressive heating and the energy released in the reaction serves to drive the compression. Deflagrations on the other hand are low-velocity expansion waves in which chemical reaction is brought about by the diffusion of heat and mass.

Equation 3.4 also shows that the pressure decrease observed in deflagrations demands an increase in velocity and the burnt gas leaving the flame has a higher velocity, a lower pressure and a lower density than the initial reactant gas. When the release of chemical energy is taken into account, with the consequent rise in temperature, it is found that the pressure change across the flame is so low that it is usually unimportant, apart from being involved in certain distortion effects (see below).

If we consider the detailed structure of the flame, it is apparent that the temperature must increase smoothly from the initial to the final state. The product concentration will change in a similar fashion, while the concentration of fuel molecules must show a corresponding decrease. These profiles are illustrated in Fig. 3.2, together with an indication of how the concentrations of intermediates are likely to vary during combustion. The visible part of the flame is located in the reaction zone and the emission is largely due to excited radicals such as CH, C_2, CHO, NH and NH_2 returning to their ground state.

An element of the flowing gas can receive heat in two ways: either from chemical reactions occurring within it or by conduction from the hotter gas ahead. Consideration of these processes along the profiles

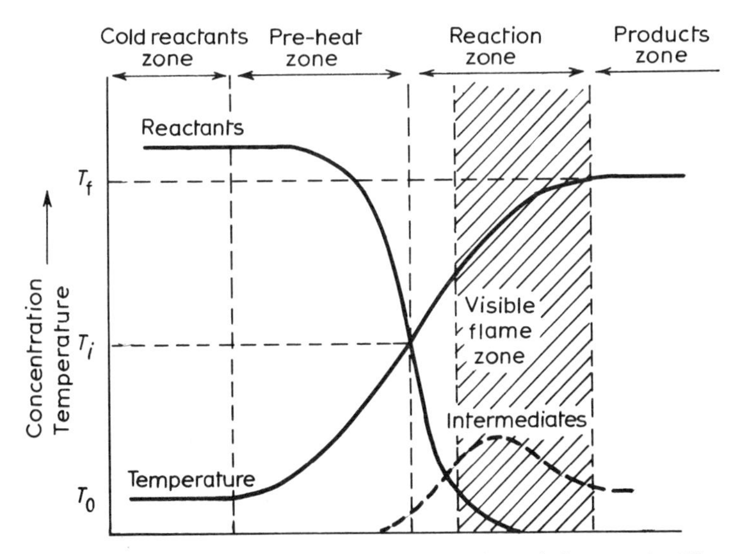

Fig. 3.2 Concentration and temperature profiles through the pre-mixed flame.

shows that two distinct regions can be recognized, separated by a point of inflexion in the temperature profile. Commencing at low temperatures it can be seen that, for any given cross-section, the heat flow into the region due to conduction is greater than the corresponding heat loss because the gradient is steeper on the high-temperature side. Beyond the point of inflexion, the converse will be true, that is, the heat loss exceeds heat gain. However, at this higher temperature the reaction rate has increased sufficiently for a significant amount of heat to be produced by chemical reaction. The temperature therefore continues to increase through the flame although at a progressively slower rate, and eventually reaches a constant value when all the fuel has been consumed and reaction has ceased. Parallel behaviour also occurs in the concentration profile, the major loss in the first region being diffusion of fuel into the flame and in the second region being consumption by chemical reaction. The first region is termed the preheating zone and the second, the reaction zone.

The two properties of the typical laminar pre-mixed flame which can be defined and, at least in principle, measured, are the *burning velocity**, S_u, and the *adiabatic flame temperature:* some examples are listed in Table 3.1. The burning velocity of a given fuel–oxidant mixture is defined as the velocity with which a plane flame front moves normal to its surface through the adjacent unburnt gas. Accurate measurement is difficult [9,48] but values are generally between 0.1 and 1.0 $m s^{-1}$, and show a small dependence on the pressure and temperature of the reactant gases, usually increasing at reduced pressures or elevated temperatures. On the other hand, the fuel–oxidant ratio has a marked effect on the burning velocity which has its maximum value with a fuel-rich mixture.

Flame temperatures usually lie in the range 2000–3500 K (Table 3.1), although flames have been generated with extremely low final temperatures even down to 35 K above the initial temperature. At the upper end of the scale, the cyanogen–oxygen system has a flame temperature of 5000 K which is associated with the stability of the products towards dissociation rather than to an excessive heat of reaction. Flame temperatures are notoriously difficult to determine and in many cases the quality of the thermodynamic data available is sufficiently high for calculations of the adiabatic flame temperature to be more reliable

* The term 'flame speed' is used by some authors for this quantity. However, it seems preferable to reserve this name for the speed of a non-stationary flame through an initially quiescent mixture. The observed flame speed in such cases includes a component due to movement of gas ahead of the flame.

Table 3.1 Flammability limits and flame properties for some common gases at atmospheric pressure. The flammability limits (after Coward and Jones [47]) apply to upward propagation in tubes. The flame temperatures refer to stoichiometric mixtures, except those marked * which are maximum values.

	Flammability limits (% by volume)		Flame temperature	Maximum burning velocity
Reactants	Lower	Upper	(K)	($m\,s^{-1}$)
$H_2 + O_2$	4.0	94	3083	11.0
$CO + O_2(+H_2O)$	15.5	94	2973	1.08
$CH_4 + O_2$	5.1	61	3010	4.5
$C_2H_2 + O_2$			3431	11.4
H_2 + air	4.0	75	2380	3.1
CO + air	12.5	74	2400	0.45
CH_4 + air	5.3	15	2222	0.45
C_2H_2 + air	2.5	80	2513	1.58
C_2H_4 + air	3.1	32	2375	0.75
C_2H_6 + air	3.1	15	2244*	0.40
C_3H_8 + air	2.2	9.5	2250*	0.43
n-C_4H_{10} + air	1.9	8.5	2256*	0.38
C_6H_6 + air	1.5	7.5	2365*	0.41
C_2H_4O + air	3.0	80	2411*	1.05

than experimental measurements. As explained in Chapter 1, the procedure for relatively cool flames is quite simple, merely requiring a calculation of the enthalpy released by conversion of reactants to products at room temperature followed by a determination of the state these products will reach when this energy is used to raise their temperature. At higher temperatures, a substantial fraction of this energy is taken up in dissociation of products and the calculations become considerably more tedious because the temperature dependence of all significant equilibria must be taken into account [9].

3.1.2 Flammability limits

Flames can only occur in mixtures within a certain composition range bounded by the limits of flammability. (The term 'inflammability' is still occasionally used but it is to be avoided because of the confusion it may cause.) These limits are usually expressed as percentages of fuel by volume (Table 3.1). There are enormous variations between fuels; thus, for acetylene the range of flammability in air is 2.5 to 80% while for propane the range is 2.2 to 9.5%. On the whole, the lower limit shows less variation; for many fuels it is about one-half the stoichiometric composition and takes the same value in both air and oxygen. As progressive amounts of inert gas are added the limits approach each other and eventually coincide (Fig. 3.3).

The burning velocity has a maximum fairly close to the stoichiometric composition and falls as the flammability limits are approached. In some systems the limits seem to correlate with a minimum flame temperature (*ca.* 1400 K for methane), the burning velocity remaining small, but finite, up to the point of extinction.

The existence of flammability limits is predicted by theories of flame propagation, provided heat loss from the burnt gas is included [49]. The heat loss may occur by conduction, convection and/or radiation and cannot normally be handled by a one-dimensional treatment so that the mathematical analysis becomes very complex.

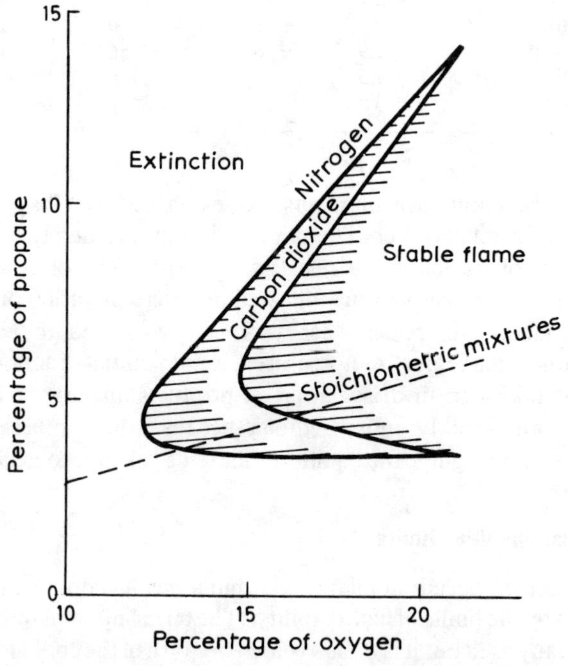

Fig. 3.3 Flammability limits for propane–air mixtures in the presence of added nitrogen or carbon dioxide (after Coward and Jones [47]).

3.1.3 Quenching

A related phenomenon is that of quenching. Close to a solid surface, heat losses are sufficient to extinguish the flame. Although loss of active species may also be involved in quenching, the nature of the surface does not appear to affect its quenching properties and it seems that quenching is primarily due to heat loss; differences in thermal conductivity between various surfaces are unimportant because the heat capacities of solids are high compared with those of gases.

The quenching diameter, d_T, of a particular gas mixture is the minimum diameter of tube through which a flame in the stationary gas mixture can propagate. The quenching distance, d_Q, is a related quantity and refers to flame propagation between parallel plates; the two quantities are related by the expression $d_T = 1.54d_Q$ [50]. Values of d_Q for some mixtures are given in Table 3.2.

Various theories of flame quenching lead to the conclusion that the dimensionless group (a Peclet number) $d_Q S_u/a$ should be constant; in this expression a is the thermal diffusivity of the unburnt gas.

A relationship in this form is to be expected on general grounds if quenching is due to removal of radicals or to heat removal. Diffusion and heat transfer are both inversely proportional to pressure, while for most flames the burning velocity is largely independent of pressure; consequently quenching distance is approximately inversely proportional to pressure. Likewise, fast-burning flames are associated with

Table 3.2 Quenching distances for various stoichiometric flames at 1 atm and room temperature [50].

System	d_Q(mm)
H_2–O_2	0.2
CH_4–O_2	0.3
C_2H_4–O_2	0.2
C_2H_2–O_2	0.1
C_3H_8–O_2	0.25
H_2–air	0.6
CH_4–air	2.5
C_2H_4–air	1.25
C_2H_2–air	0.5
C_3H_8–air	2.1
C_6H_6–air	1.9
i-C_8H_{18}–air	2.6

small quenching distances. This result has been confirmed by experiment and values of $d_Q S_u/a$ are generally between 40 and 50 [51].

Quenching of flames is important in flame traps. These are devices used to prevent flame propagation: the example most often quoted is the miner's safety lamp in which the flame is surrounded by a fine copper gauze. It will be evident from the discussion above that the mesh size is crucial in determining whether the passage of a particular flame will be stopped and careful design is necessary if the device is to be safe in all conditions which might be encountered. In industrial applications, flame traps frequently comprise an assembly of narrow-bore, thin-walled tubes which have a minimal resistance to gas flow while preventing flame propagation.

3.1.4 Stabilization of flames on burners

Since a combustion wave travels at a characteristic burning velocity, a stationary flame may be obtained by flowing the pre-mixed gases at the same speed in the reverse direction. Such a flame would possess only neutral stability and its position would shift in an uncontrolled manner. In household and industrial appliances, flame stability is achieved by attaching the flame to a simple device known as a burner.

For a typical laminar pre-mixed flame, the burner fulfils three functions: it contains an arrangement for mixing fuel and oxidant in the appropriate proportions, it provides a suitably shaped section to provide laminar flow and it acts as a heat sink which restricts movement of the flame. The simplest example of such a device is the Bunsen burner.

The actual stabilizing effect of the burner is provided by the rim alone and a simple metal ring will perform the same function. The effect of the rim is to remove heat (and possibly active species) from the flame and hence to reduce the burning velocity in its vicinity. If a combustion wave is situated immediately above the burner rim, then its position will be determined by the relative magnitudes of the burning velocity and the flow velocity. If the burning velocity is greater than the flow velocity, the flame will move downwards until the burning velocity decreases to equal the flow velocity. Alternatively, if the flow velocity exceeds the burning velocity, the flame will lift until the two become equal. Thus, within certain flow velocity limits, the flame will be held in place above the rim.

For a laminar flow burner, the flow velocity is very low close to the walls and above the rim but increases towards the centre of the burner

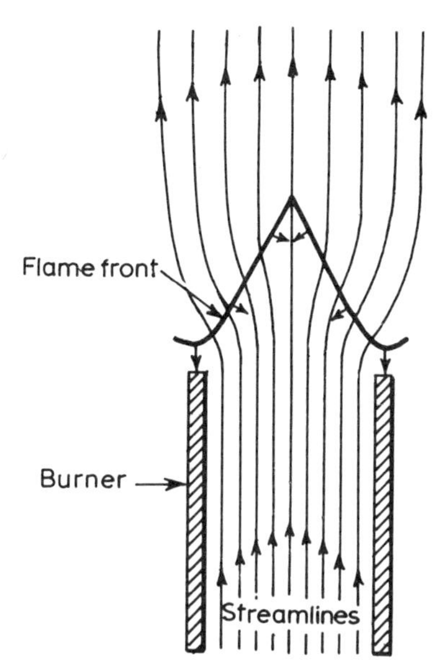

Fig. 3.4 Streamlines through the laminar, pre-mixed burner flame. The length of the arrows on the flame front indicates the relative magnitude of the burning velocity.

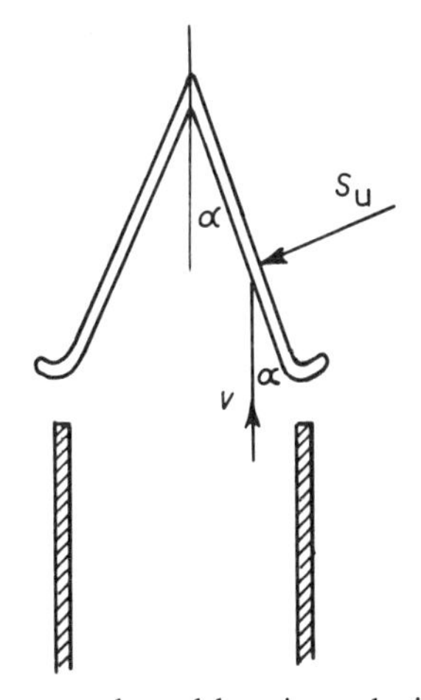

Fig. 3.5 Flame cone angle and burning velocity, $S_u = v \sin \alpha$.

giving a parabolic velocity profile. At all points within the rim but outside the quenching distance, the flow velocity exceeds the burning velocity and the flame is inclined upwards from the rim so that the burning velocity equals the component of the flow velocity normal to the flame front. This accounts for the familiar conical appearance of the flame (Fig. 3.4). Measurement of the gas flow velocity and the cone angle using a nozzle designed to give a flat velocity profile forms the basis of one method of measuring the burning velocity (Fig. 3.5).

The flame is only stabilized on the burner within certain flow velocity limits. If the gas flow is progressively reduced a point will eventually be reached at which the burning velocity exceeds the gas velocity somewhere across the burner diameter. At the *flash-back limit*, the flame will become unstable and will propagate back down the burner tube. The situation then inside the burner tube is represented in Fig. 3.6 for various gas flow velocity gradients. If the gas velocity is represented by line a, then the flow velocity exceeds the burning velocity and the flame will rise out of the tube until it can stabilize by adopting the normal conical shape. Line c allows the burning velocity to exceed the

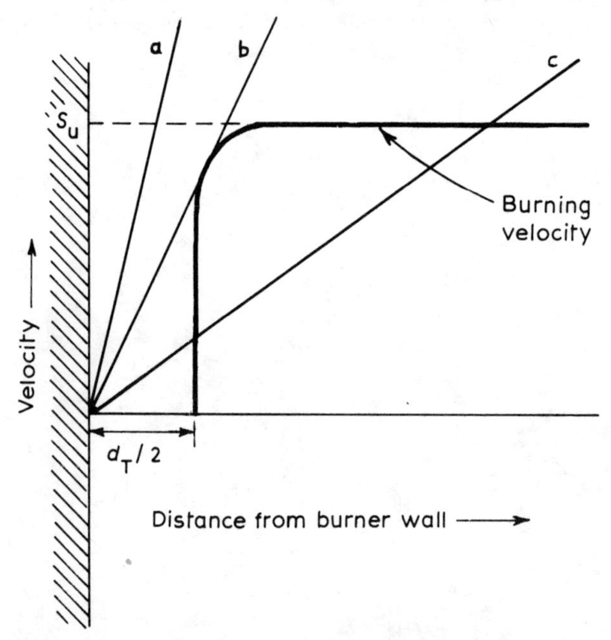

Fig. 3.6 Burning velocity and gas velocity inside a burner tube (after Lewis and Von Elbe [32]).

flow velocity and the flame will flash back. The critical gradient b corresponds to the flash-back limit and is denoted by g_F: it will be seen from Fig. 3.6 that g_F is approximately equal to $2S_u/d_T$.

A little above the flash-back limit, a tilted flame may occur. The back pressure of the flame has allowed distortion of the flow and in the region where the flow velocity is reduced the flame will enter the burner. Because of the constraint provided by the burner tube, the flow there is less liable to distortion so that further propagation is prevented. The stable distorted, or tilted, flame illustrated in Fig. 3.7 then remains.

Towards the edge of the burner the flame characteristics will be affected by the entrainment of atmospheric air. This is not particularly important when the flame lies close to the rim but, as the flow velocity increases, the flame will rise in order to allow the burning velocity to increase. The dilution by atmospheric air has the opposite effect, causing the burning velocity to fall. The flame will therefore continue to rise and eventually become unstable at the *blow-off limit*. This limit is also characterized by a critical velocity gradient, now denoted by g_B. In practice, g_F and g_B are defined by the values of $8\bar{u}/d$ at which the limits occur, $\bar{u}$ being the mean flow velocity in the burner at the limit, that is,

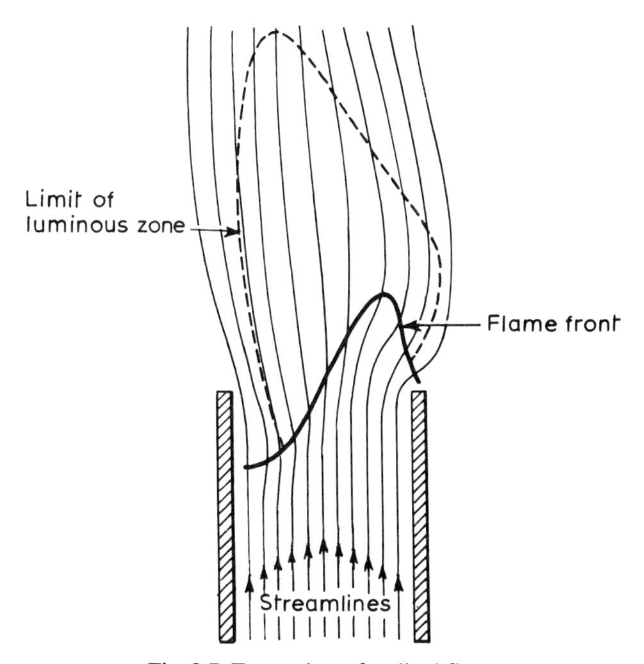

Fig. 3.7 Formation of a tilted flame.

the volume flow rate divided by the cross-sectional area, and d the diameter of the burner tube.

In general, the values of the gas velocities at the two limits will depend on burner dimensions and on the gas composition. For fuel-rich mixtures and high gas velocities, a second phenomenon may occur. Due to the entrainment of atmospheric air, the mixture will become progressively leaner above the burner eventually approaching the stoichiometric composition. Because of the increased burning velocity, a lifted flame is able to form some distance above the burner.

The lifted flame displays two stability limits similar to those of the seated flame. When the gas velocity is reduced, drop-back occurs, the flame taking up its normal position on the burner. At high velocities the flame will blow-out. Fig. 3.8 illustrates schematically the velocity–composition regions within which each flame is stable: note that a situation exists where either flame may occur.

Due to the phenomenon of quenching described in the previous section, a flame cannot be supported on a burner whose dimensions fall below a certain size, the quenching diameter, which depends on the gas

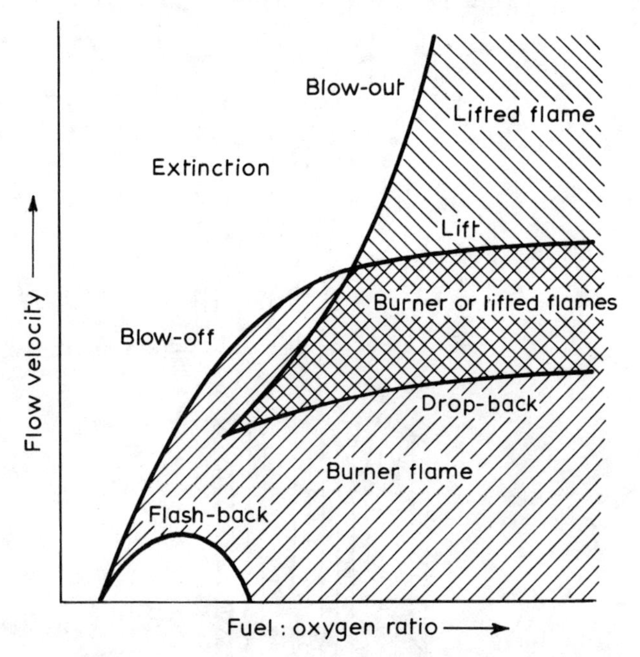

Fig. 3.8 Schematic representation of limiting conditions for various pre-mixed flames.

composition and pressure. Because this critical diameter is approximately proportional to the reciprocal of the pressure, its value is most easily determined by measuring the pressure at which the flame is extinguished for given sizes of burner. The flow velocity limits for flash-back and blow-off also coincide when quenching occurs.

Burner design

It is worthwhile at this stage considering some of the factors involved in the design of practical burners for both industrial and domestic applications. Clearly the tube diameter must exceed the quenching distance. For a stable cone to form, it is also found that the gas flow velocity must lie between $2S_u$ and $5S_u$. The flash-back and blow-off behaviour then impose upper and lower limits on the velocity:diameter ratio. Finally, flow up the burner tube must be laminar and therefore the Reynolds number ($Re = du\rho/\eta$) must not exceed 2300. These constraints are summarized graphically in Fig. 3.9 where the hatched area indicates the relatively restricted velocity–diameter regime in which a burner can function.

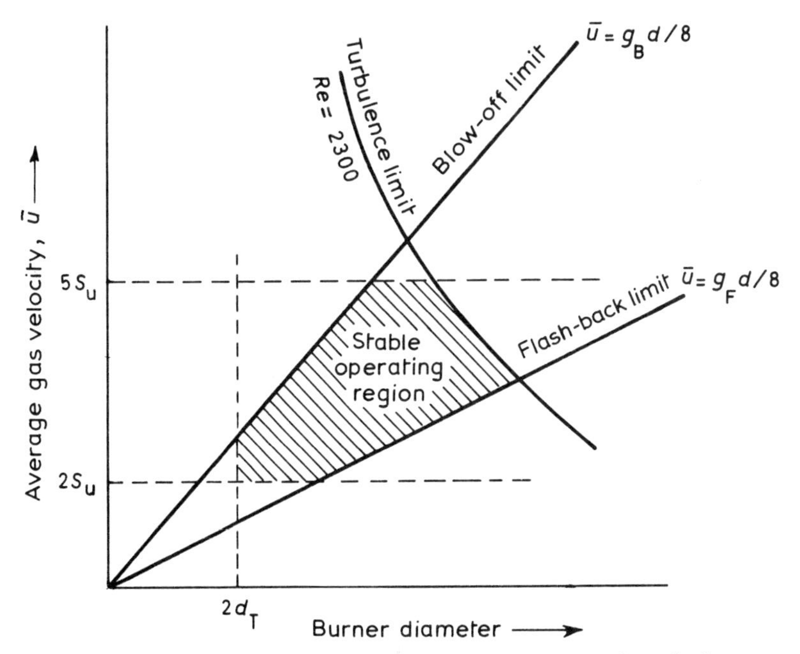

Fig. 3.9 Parameters determining the region for stable operation of a bunsen burner (after Glassman [52]).

The pre-mixed gas composition is usually expressed in terms of an equivalence ratio, Φ, which is the actual fuel: oxidizer ratio divided by the fuel: oxidizer ratio corresponding to complete combustion to carbon dioxide and water; the latter ratio is often referred to as the stoichiometric fuel: oxidizer ratio. The dependence of g_F and g_B on Φ is illustrated in Fig. 3.10.

The equivalence ratio is controlled by the ratio of the burner tube area to the fuel orifice area. For obvious reasons, the fuel orifice area is kept below the quenching diameter. The momentum of the fuel entering through an orifice area A_F at a velocity u_F is $A_F \rho_F u_F^2$. Entrainment of air then occurs so that momentum is conserved

$$A_B \rho_M u_M^2 = A_F \rho_F u_F^2 \tag{3.5}$$

where A_B is the area of the burner tube and the subscript M refers to the mixture. The volumetric fuel: mixture ratio

$$\frac{[F]}{[M]} = \frac{A_F u_F}{A_B u_M} \tag{3.6}$$

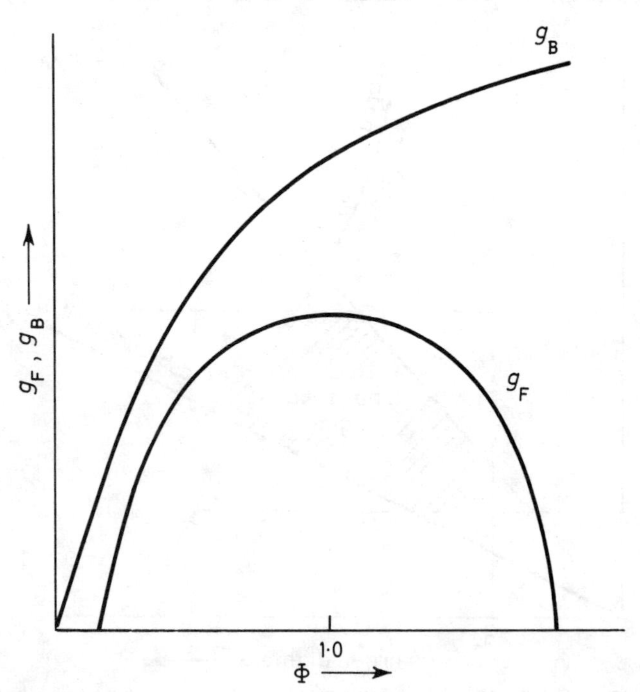

Fig. 3.10 Blow-off and flash-back limits as a function of the equivalence ratio, Φ.

and thus

$$\frac{[F]}{[M]} = \left(\frac{\rho_M}{\alpha \rho_F}\right)^{1/2} \tag{3.7}$$

where $\alpha = A_B/A_F$
A corresponding, but much more cumbersome expression for the volumetric fuel: air ratio in terms of ρ_A/ρ_F and α can be derived from Equation 3.7, if necessary.

The particular application of Equation 3.7 is in establishing how a burner must be changed if it is to handle a different fuel, as happened in the United Kingdom when burners operating on town gas, which was primarily a mixture of hydrogen and carbon monoxide, were converted to use natural gas. The first requirement in such a situation is to select the appropriate fuel orifice in order to maintain the correct equivalence ratio and then to adjust the fuel supply pressure to give approximately the same heat output while remaining within the operating regime shown in Fig. 3.9.

3.1.5 Mechanism of flame propagation

Of the two fundamental characteristics of a flame, burning velocity and flame temperature, the calculation of the latter has already been outlined. Theories of flame propagation attempt to predict the numerical value of the former quantity from knowledge of the physical and chemical properties of the materials involved.

Theories generally assume a temperature profile through a pre-mixed flame as illustrated in Fig. 3.11. The point of inflexion is termed the ignition point and it is convenient to locate the origin of the coordinates there. The profile in the pre-heating zone is concave upwards while that in the reaction zone is concave downwards. The ignition point coincides approximately with the onset of chemical reaction.

It was first thought that the temperature T_i at the ignition point (sometimes loosely termed the ignition temperature) must be related to the self-ignition temperature, T_{ign}, at which spontaneous ignition occurs. In fact, experiments demonstrate that the two are quite unconnected so that, for example, the effect of trace additives on one of these temperatures bears no relation to their effect on the other. Reaction at the self-ignition temperature builds up only slowly during an appreciable induction period (up to 1s) whilst the residence time of the fuel in the reaction zone is so short (typically 10^{-5} s) that reaction

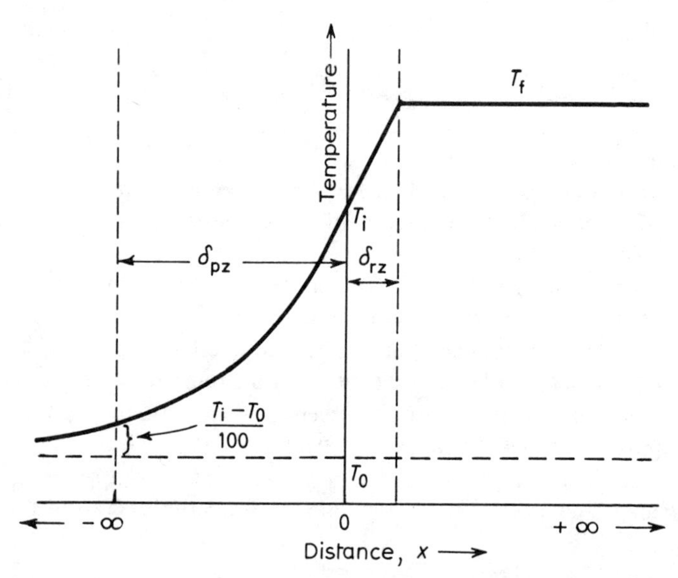

Fig. 3.11 Elementary model of the temperature profile through a pre-mixed flame.

after the ignition point must be extremely fast. In consequence, if flame propagation is primarily thermal in nature, the temperature at the ignition point must be quite close to the final flame temperature. This need not be the case, of course, if diffusion of active intermediates is responsible for the onset of combustion.

(a) *The Mallard–Le Chatelier Model*

The first significant attempt at solving the problem was made by Mallard and Le Chatelier in 1883 [53]. While very much over-simplified, it still provides a useful basis for understanding the phenomenon.

First, it is assumed that the temperature gradient in the reaction zone is linear and equal to $(T_f - T_i)/\delta_{rz}$ where δ_{rz} is the *reaction zone length.* The heat flux into the pre-heating zone is then given by $\lambda(T_f - T_i)/\delta_{rz}$ where λ is the thermal conductivity. This must equal the rate of increase of enthalpy in this region, given by $\dot{m}\bar{c}_p(T_i - T_0)$ where $\dot{m}$ is the mass flow rate and $\bar{c}_p$ the mean specific heat capacity. For a one-dimensional system of uniform cross-section, $\dot{m}$ is simply equal to the product of flow velocity and density, $u\rho$.

The total reaction time is equal to δ_{rz}/u and hence a mean reaction rate, $\bar{J}'$, may be defined as its reciprocal, that is, $\bar{J}' = u/\delta_{rz}$.

Combining these expressions gives

$$u^2 = \frac{\lambda}{\rho\bar{c}_p}\frac{(T_f - T_i)\bar{J}'}{(T_i - T_0)}$$

or, the burning velocity,

$$S_u = \left[\frac{\lambda}{\rho\bar{c}_p}\frac{(T_f - T_i)\bar{J}'}{(T_i - T_0)}\right]^{1/2} \tag{3.8}$$

Since $\lambda/\rho\bar{c}_p = a$, the thermal diffusivity, this simple treatment leads to the conclusion, confirmed by more advanced theories of flame propagation, that

$$S_u \propto (a\bar{J}')^{1/2} \tag{3.9}$$

While it does not permit an accurate calculation of burning velocity, it nevertheless enables the effect of changing parameters such as temperature and pressure to be predicted with some precision.

(b) *Diffusional propagation of flames*

Mallard and Le Chatelier [53] assumed that flame propagation was due entirely to heat conduction. It was subsequently appreciated that the diffusion of active species ahead of the flame could provide an equally plausible explanation of events and theoretical treatments became subdivided into either thermal or diffusional theories according to the convictions of their protagonists. Attempts to resolve the controversy by experiment were little help since both processes obey similar differential equations and hence respond in much the same way to external stimuli.

A mechanism based purely on diffusion carries with it the assumption of isothermal propagation. While clearly of limited applicability, the approach still merits a brief review. In the best-known theory, due to Tanford and Pease [54], it is assumed that thermodynamic equilibrium is attained at the end of the reaction zone. The active centres then diffuse upstream into the reaction zone where their concentrations are in excess of the equilibrium values for the appropriate temperature. A mean temperature for the reaction zone, lower than the final flame temperature, is assumed so that constant values for the density and the flow velocity are obtained. A concentration profile (Fig. 3.12) can then be found for the active centres by solving the appropriate diffusion equation, together with a small correction term for the loss of active centres by reaction. As the reaction rate depends

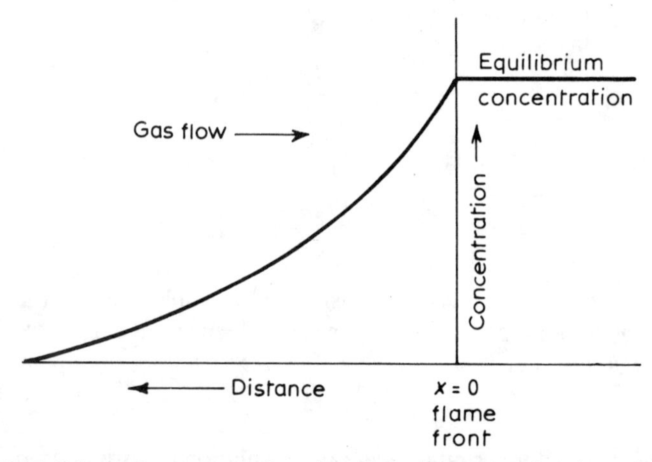

Fig. 3.12 Schematic representation of the radical concentration through a premixed flame (after Tanford and Pease [54]).

mainly on the concentration of active centres and less on temperature, the assumption at this stage of a constant temperature through the reaction zone does not introduce any substantial error. Using this concentration profile the calculation is then completed by integrating the rate expression through the reaction zone and equating the result to the total loss of reactants. A more correct temperature profile may be obtained by solving the heat transfer equation assuming the major loss of energy occurs only when the final flame temperature is attained. The following relation is obtained for the burning velocity,

$$S_u^2 = \sum_j \frac{k_j[\mathrm{F}]y_j D_{j0}}{y' B_j} \tag{3.10}$$

where k_j is the rate constant for reaction of the j-th species at the final flame temperature, y_j is the mole fraction of species j in the burnt gas, [F] is the concentration of fuel, D_{j0} is the diffusion coefficient of species j in the unburnt gas, y' is the mole fraction of the combustion product (i.e. fuel or oxidant, whichever is less than stoichiometric), and B_j is a factor of order unity which corrects for the loss of active centres by chemical processes within the reaction zone.

An important deduction from this expression is that S_u is approximately proportional to $(\sum y_j D_{j0})^{1/2}$, which explains the observed dependence of burning velocity on concentration of active centres.

Theories of this sort take no account of the branching-chain nature of most combustion reactions. The analysis given above assumed that the reaction rate increases continuously through the flame and therefore that reaction occurs an infinite distance ahead of the flame front. For a branching-chain reaction, the rate is negligibly slow until the rate of branching exceeds that of chain breaking after which the reaction becomes explosive.

(c) *Comprehensive theory*

The advent of powerful computers means that it is no longer necessary for theories of flame propagation to ignore diffusion of radicals in a thermal theory or heat transfer in a diffusional theory; nor is it necessary to use simplified global kinctics. It is now possible to solve numerically a complicated set of simultaneous partial differential equations and it has therefore become the custom to include all heat transport and diffusion processes, as well as the detailed reaction chemistry. Because of the close coupling between these two phenomena, it has always been questioned whether attempting to separate them is really meaningful. However, it seems clear that heat conduction is fundamental to the propagation of pre-mixed flames but that diffusion may add an extra contribution, particularly when low molecular weight species, such as hydrogen atoms, are involved.

When the chemistry of the flame reactions is fully understood and accurate data are available, complete computer calculations give extremely close agreement with experimental burning velocities: for example, a calculated value of 0.289 m s^{-1} for an H_2–40%Br_2 flame at an initial temperature of 323 K may be compared with an experimental measurement of 0.29 m s^{-1} [55]. Indeed, the agreement nowadays is usually well within the uncertainties in the reaction kinetic data. Measurements of burning velocities and of concentration profiles through the flame are therefore used in conjunction with computer simulations to obtain more accurate values of the rate constants involved [56–58].

In the following treatment we shall see how the complete problem is set up for computer solution and will then introduce a number of simplifying assumptions in order to achieve an analytical solution. In doing so, we shall explicitly reduce the treatment to the classical thermal model and the final conclusions will prove only marginally more significant than those of the elementary derivation already given. However, the exercise is valuable because the same approach is used to solve other combustion phenomena, for example, the burning of droplets.

The theory makes use of the equations for conservation of mass

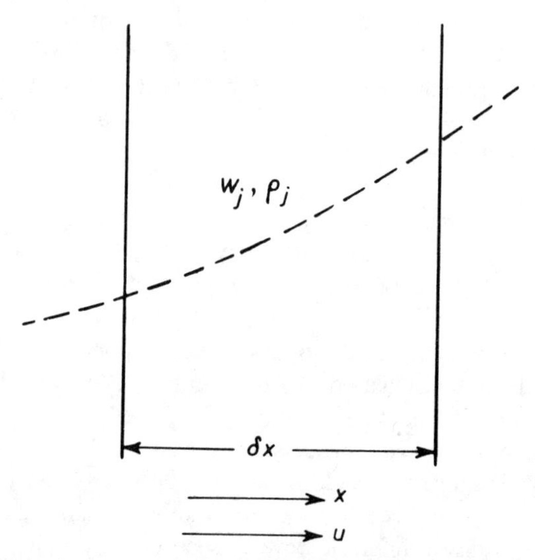

Fig. 3.13 Notation used in the mathematical model of a laminar flame.

(continuity) and for conservation of energy and assumes that flow is one-dimensional, with pressure constant throughout. The derivation is based on a thin volume element of unit area and thickness δx, with space-fixed co-ordinates normal to the gas flow which is from left to right (Fig. 3.13). We consider first a chemical species j whose density ρ_j and mass fraction $w_j = \rho_j/\rho$ increase from left to right.

Continuity demands that the total mass flow is the same at each boundary, although the species (convective) flow will differ because of the change in mass fraction through the volume. The net change in unit time is then given by the difference between the mass flow in at the left-hand boundary and out at the right-hand boundary, that is,

$$\rho u w_j - \rho u \left(w_j + \frac{\mathrm{d}w_j}{\mathrm{d}x} \delta x \right)$$

which may be simplified to

$$-\rho u \frac{\mathrm{d}w_j}{\mathrm{d}x} \delta x$$

Diffusion flow, due to the concentration gradient, follows Fick's Law and hence is in the opposite direction, thus

$$-D_j \frac{\mathrm{d}(\rho w_j)}{\mathrm{d}x} + \left\{ D_j \frac{\mathrm{d}(\rho w_j)}{\mathrm{d}x} + \frac{\mathrm{d}}{\mathrm{d}x}\left[D_j \frac{\mathrm{d}(\rho w_j)}{\mathrm{d}x} \right] \delta x \right\}$$

which is equal to

$$\frac{d}{dx}\left[D_j \frac{d(\rho w_j)}{dx}\right]\delta x$$

Different chemical species have different molecular velocities because of their different masses and these velocities in their turn are temperature dependent so that diffusion will also occur under the influence of a temperature gradient. This process of thermal diffusion is known to have only a small effect in most cases and has been neglected in the present treatment; it is, however, included in more detailed analyses [56–58].

The remaining term is due to chemical change and is written here as a loss by chemical reaction $J_j\delta x$ where J_j is the rate expressed as density change in unit time (kg m^{-3} s^{-1}). The total change in unit time is then given by

$$\frac{d\rho_j}{dt}\delta x = \frac{d}{dx}\left[D_j \frac{d(\rho w_j)}{dx}\right]\delta x - \rho u \frac{dw_j}{dx}\delta x - J_j \delta x \qquad (3.11)$$

Dividing through by δx and considering only the steady-state solution for which $d\rho_j/dt = 0$ leads to

$$\frac{d}{dx}\left[D_j \rho \frac{dw_j}{dx}\right] - \rho u \frac{dw_j}{dx} - J_j = 0 \qquad (3.12)$$

An equation of this form can be written for each of the chemical species present in the system.

We can now perform exactly the same task for the enthalpy, thereby expressing in mathematical terms the conservation of energy. Each of the terms in Equation 3.12 is replaced as follows:

Diffusion flux $\frac{d}{dx}\left[D_j \rho \frac{dw_j}{dx}\right]$ by heat conduction $\frac{d}{dx}\left(\lambda \frac{dT}{dx}\right)$

Convective mass flux $\rho u \frac{dw_j}{dx}$ by convective energy flow $\rho u c_p \frac{dT}{dx}$

Species loss by reaction J_j by enthalpy loss $\sum(-h_j J_j)$

The differential flow of species with different heat capacities has been neglected, as have contributions from viscous forces, changes in kinetic energy and radiative transfer. All these terms, while minor, can be included if the problem is to be solved numerically. Without them, the energy equation becomes

$$\frac{\mathrm{d}}{\mathrm{d}x}\left(\lambda\frac{\mathrm{d}T}{\mathrm{d}x}\right)-\rho uc_p\frac{\mathrm{d}T}{\mathrm{d}x}+\sum h_jJ_j=0 \tag{3.13}$$

We are left with a set of $(j+1)$ differential equations which can only be solved numerically. To achieve an analytical solution further simplification is required.

Most flames involve only two initial reactants and we can begin by regarding the chemical reaction as involving only two species, a fuel denoted by the subscript F and an oxidant denoted by OX, so that the only species conservation equations of interest are

$$\frac{\mathrm{d}}{\mathrm{d}x}\left(\rho D_{\mathrm{F}}\frac{\mathrm{d}w_{\mathrm{F}}}{\mathrm{d}x}\right)-\rho u\frac{\mathrm{d}w_{\mathrm{F}}}{\mathrm{d}x}-J_{\mathrm{F}}=0 \tag{3.14}$$

$$\frac{\mathrm{d}}{\mathrm{d}x}\left(\rho D_{\mathrm{OX}}\frac{\mathrm{d}w_{\mathrm{OX}}}{\mathrm{d}x}\right)-\rho u\frac{\mathrm{d}w_{\mathrm{OX}}}{\mathrm{d}x}-J_{\mathrm{OX}}=0 \tag{3.15}$$

In making this approximation we are disregarding the separate diffusion of active species and, in that sense, we are adopting a thermal theory although diffusion is still allowed to occur. We make a further assumption that $D_{\mathrm{F}}=D_{\mathrm{OX}}=D$, which is reasonable if the molecular weights are not too dissimilar, and also note that the chemical rates must be linked by the stoichiometric coefficient, s, that is, $J_{\mathrm{F}}=J_{\mathrm{OX}}/s$.

Dividing Equation 3.15 by s and subtracting the result from Equation 3.14 gives

$$\frac{\mathrm{d}}{\mathrm{d}x}\left[\rho D\frac{\mathrm{d}}{\mathrm{d}x}(w_{\mathrm{F}}-w_{\mathrm{OX}}/s)\right]-\rho u\frac{\mathrm{d}}{\mathrm{d}x}(w_{\mathrm{F}}-w_{\mathrm{OX}}/s)=0 \tag{3.16}$$

The importance of this equation is that it no longer contains a source term and the quantity $(w_{\mathrm{F}}-w_{\mathrm{OX}}/s)$ therefore behaves like the mass fraction of an inert species: it is a *conserved property* and remains constant throughout. Identifying such conserved properties proves to be of great importance in solving combustion problems [59].

It was pointed out in Chapter 1 that the enthalpies of individual species are assigned with respect to an arbitrary zero at an appropriate reference temperature. In the present case, the only constraint on the enthalpies is that they obey the relation

$$[(s+1)h_{\mathrm{PROD}}]-[h_{\mathrm{F}}+sh_{\mathrm{OX}}]=-q \tag{3.17}$$

where h is the enthalpy per unit mass of substance and q is the heat released by combustion of unit mass of fuel. It is therefore convenient to choose

$$h_{\mathrm{OX}} = h_{\mathrm{PROD}} = h_{\mathrm{INERT}} = 0$$

and

$$h_{\mathrm{F}} = q \tag{3.18}$$

at the unburnt gas temperature, T_0. Assuming that the heat capacity remains unchanged enables the total enthalpy to be written as

$$\begin{aligned} h &= c_p(T - T_0) + w_{\mathrm{F}} h_{\mathrm{F}} \\ &= c_p(T - T_0) + w_{\mathrm{F}} q \end{aligned} \tag{3.19}$$

Now multiplying Equation 3.14 by q gives

$$\frac{\mathrm{d}}{\mathrm{d}x}\left[\rho D \frac{\mathrm{d}(q w_{\mathrm{F}})}{\mathrm{d}x}\right] - \rho u \frac{\mathrm{d}}{\mathrm{d}x}(q w_{\mathrm{F}}) - q J_{\mathrm{F}} = 0 \tag{3.20}$$

Then, in Equation 3.13 we may replace T by $(T - T_0)$ and, using Equation 3.18, replace $\sum h_j J_j$ by $q J_{\mathrm{F}}$ giving

$$\frac{\mathrm{d}}{\mathrm{d}x}\left\{\frac{\lambda}{c_p} \frac{\mathrm{d}[c_p(T - T_0)]}{\mathrm{d}x}\right\} - \rho u \frac{\mathrm{d}}{\mathrm{d}x}[c_p(T - T_0)] + q J_{\mathrm{F}} = 0 \tag{3.21}$$

For a Lewis Number of unity, $\lambda / c_p = \rho D$, and Equations 3.20 and 3.21 can be added

$$\frac{\mathrm{d}}{\mathrm{d}x}\left\{\rho D \frac{\mathrm{d}}{\mathrm{d}x}[c_p(T - T_0) + q w_{\mathrm{F}}]\right\} - \rho u \frac{\mathrm{d}}{\mathrm{d}x}[c_p(T - T_0) + q w_{\mathrm{F}}] = 0 \tag{3.22}$$

and using Equation 3.19 this becomes

$$\frac{\mathrm{d}}{\mathrm{d}x}\left(\rho D \frac{\mathrm{d}h}{\mathrm{d}x}\right) - \rho u \frac{\mathrm{d}h}{\mathrm{d}x} = 0 \tag{3.23}$$

Thus, the enthalpy defined as above is also a conserved property and the sum of the chemical and thermal energies remains constant throughout.

The purpose of this somewhat lengthy analysis is to show that, by making various approximations, it is possible to change the species conservation and energy conservation equations into two new equations which are identical in form. This means that it is only necessary to solve a single differential equation: for convenience we shall choose to solve the energy equation, Equation 3.13.

(i) *The pre-heating zone*

As explained previously, the rate of reaction in the pre-heating zone is negligible and the temperature is determined by the balance between the convective and diffusive terms, so that the energy equation reduces to

$$\lambda \frac{d^2T}{dx^2} - \rho u c_p \frac{dT}{dx} = 0 \tag{3.24}$$

Thus for any volume element in this region, heat is received by conduction from right to left and lost by convective (mass) flow from left to right, since the gas is hotter when it leaves.

From Fig. 3.11 the appropriate boundary conditions for the pre-heating zone are

$$x = -\infty, T = T_0, dT/dx = 0$$
$$x = 0, T = T_i$$

and Equation 3.24 may be solved immediately to give

$$T - T_0 = (T_i - T_0)\exp(\rho u c_p x/\lambda) \tag{3.25}$$

To the extent that c_p and λ can be adequately represented by mean values, the temperature increases exponentially with distance in the pre-heating zone. Although in principle the temperature starts to rise from $x = -\infty$, a thickness may be arbitrarily defined by considering the zone to commence at $(T - T_0)/(T_i - T_0) = 1/100$ (i.e. $\sim T_0 + 20$ K for a typical ignition point temperature of 2000 K). Replacing u by S_u gives the thickness of the pre-heating zone

$$\delta_{pz} = \frac{4.6\lambda}{c_p \rho S_u}$$

Typical values of δ_{pz} range from 10^{-5} m for a burning velocity of 30 m s^{-1} to 10^{-2} m for a burning velocity of 0.03 m s^{-1}.

The temperature gradient at the ignition point, $x = 0$ is

$$\left(\frac{dT}{dx}\right)_{x=0} = \frac{\rho u c_p (T_i - T_0)}{\lambda} \tag{3.26}$$

and this will be required in the next section.

(ii) *The reaction zone*

Above the temperature T_i, which is close to the temperature T_f, the reaction rate becomes appreciable. The temperature profile now becomes concave downwards meaning that heat loss by conduction to the left exceeds that received from the right. The net heat loss is made up by heat released in the reaction. The small convective loss term may be neglected so that the energy Equation 3.13 now becomes

$$\lambda \frac{d^2T}{dx^2} + qJ_F = 0 \tag{3.27}$$

In order to solve this equation we make use of the identity

$$\frac{d}{dx}\left(\frac{dT}{dx}\right)^2 = 2\left(\frac{dT}{dx}\right)\left(\frac{d^2T}{dx^2}\right) \tag{3.28}$$

Multiplying Equation 3.27 by $(2/\lambda)(dT/dx)$ gives

$$\frac{d}{dx}\left(\frac{dT}{dx}\right)^2 = \frac{-2qJ_F}{\lambda}\frac{dT}{dx} \tag{3.29}$$

which integrates to

$$\left(\frac{dT}{dx}\right)^2\Bigg]_{x=0}^{x=\infty} = \frac{-2q}{\lambda}\int_{T_0}^{T_f} J_F dT \tag{3.30}$$

J_F is a sensitive function of T and strictly the appropriate Arrhenius expression should be introduced into Equation 3.30. However, for the present purpose it is sufficient to note that T_i is very close to T_f, at which temperature most of the reaction occurs and the variable J_F may be replaced by $\bar{J}_F$, the temperature-averaged reaction rate. As $dT/dx = 0$ as x approaches infinity, the equation now becomes

$$\left(\frac{dT}{dx}\right)_{x=0} = \left[\frac{2q}{\lambda}\bar{J}_F(T_f - T_0)\right]^{1/2} \tag{3.31}$$

At the ignition point, the temperature gradient must have the same value irrespective of whether it is derived by solving the energy equation in the pre-heating zone or in the reaction zone. Thus, we may equate Equations 3.26 and 3.31

$$\frac{\rho u c_p(T_i - T_0)}{\lambda} = \left[\frac{2q}{\lambda}\bar{J}_F(T_f - T_0)\right]^{1/2} \tag{3.32}$$

Noting that $T_i \approx T_f$ and that the burning velocity $S_u = u$, rearrangement of Equation 3.32 yields

$$S_u = \left[\frac{2q\bar{J}_F\lambda}{\rho^2 c_p^2(T_f - T_0)}\right]^{1/2} \tag{3.33}$$

Now if $w_{F,0}$ is the mass fraction of fuel when $T = T_0$, then

$$w_{F,0}q = c_p(T_f - T_0) \tag{3.34}$$

since the enthalpy released by combustion of this fuel raises the temperature of the flame gases from T_0 to T_f. Also the thermal diffusivity $a = \lambda/\rho c_p$ and so Equation 3.33 can be written

$$S_u = \left(\frac{2a\bar{J}_F}{w_{F,0}\rho}\right)^{1/2} = \left(\frac{2a\bar{J}_F}{\rho_{F,0}}\right)^{1/2} \tag{3.35}$$

where $\rho_{F,0}$ refers to the density of the fuel when $T = T_0$.

In view of the numerous approximations made it would be unrealistic to expect this expression to give precise values of burning velocities. Nevertheless it is useful in predicting the way in which burning velocity may change with alteration in conditions such as the pressure and the presence of an inert diluent. It is also worth noting that it leads to the same result (Equation 3.9) as that obtained previously using the simple Mallard–Le Chatelier treatment and that both expressions include terms relating to heat transport and reaction kinetics.

3.1.6 Turbulent combustion

So far the discussion has been confined to streamline, laminar flow conditions in which elements of the gas flow more or less along parallel lines following the contours of the adjacent solid surface. However, if the tube diameter, or the flow velocity, is progressively increased, a stage is reached at which the flow becomes irregular or turbulent and is no longer streamlined. Turbulence may also be generated by the presence of an obstacle such as a mesh or grid in the flowing gas. Whatever the origin of the turbulence in a particular system, it has a very marked effect on the burning process, often drastically increasing the rate of flame propagation and there has been a considerable increase in interest in turbulent combustion in recent years [60].

Turbulent flow may be regarded as random local motions superimposed on the uniform motion of the fluid. Terms like *eddies*, *vortices*, *turbulent spheres* and *turbulent balls* are all employed to describe such motions. In all these cases, one visualizes a transient volume of gas, usually cylindrical or spherical in shape, whose size is characterized by a linear dimension, L, rotating at a rate characterized by a fluctuation velocity, u_{fl}. Although any system will contain a range of values of L, one can choose an appropriate average $\bar{L}$, which

provides a measure of the scale of the turbulence. The mixing length, that is the average distance turbulent spheres travel before colliding, is also roughly equal to the mean diameter $\bar{L}$. The mean of the fluctuation velocities will be zero but one can define a *root mean square fluctuation velocity*, $u' = (\overline{u_{\mathrm{fl}}^2})^{1/2}$. *Turbulence intensity* may be characterized by u'/u, where u is the main flow velocity.

Turbulent motion appears first as large stable eddies which are anisotropic. Vortex stretching leads to small, faster moving eddies which are more isotropic. Eventually the eddies reduce to the dimensions of molecular motion, the energy being dissipated by viscosity into random kinetic energy of the molecules. Because angular momentum is conserved in this cascade process, the kinetic energy involved increases. This picture of eddy decay refers only to what takes place in the free stream. At the origin of the turbulence, for example, where two gas streams of different velocity meet, the reverse occurs: the shear forces create small-scale vortices which coalesce into larger eddies.

The state of a gas flow is usually characterized by the dimensionless *Reynolds number*, *Re*, defined by

$$Re = \frac{ud\rho}{\eta} \tag{3.36}$$

where u is a flow velocity, d is a characteristic length, and η is the dynamic viscosity. If u is chosen as the average flow velocity and d as the burner diameter, then the Reynolds number so obtained will provide a criterion for the transition from laminar to turbulent flow. Below $Re = 2300$ the flow is normally laminar, while above $Re = 3200$ it becomes turbulent. In the intermediate range 2300–3200, the behaviour is not so readily predictable: small perturbations are quickly damped while more severe disturbances persist. The Reynolds number is the ratio of inertial to viscous forces in the flow. Sometimes it is convenient to define other Reynolds numbers; for example, $Re_L = u'L\rho/\eta$, the *turbulent Reynolds number* based on the length scale L. A complete description of turbulence requires a small length scale, as well as an integral length scale and in the cascade process outlined above it is possible to define a scale parameter λ (the Taylor microscale) associated with the small dissipative eddies and hence a *microscale Reynolds number*, Re_λ. This Reynolds number Re_λ, is the ratio of the time scales for large and small eddies.

We can now examine the effects of turbulent motion on combustion. When the size of the turbulent eddies does not exceed the flame front thickness, the only effect of this small-scale turbulence is to increase heat and mass transport through the flame front. It was shown in Chapter 1 that molecular transport processes involve the product of the mean molecular velocity $\bar{c}$ and the distance between molecules or mean free path, l. For transport by eddy motion, this term is replaced by the corresponding product $u'\bar{L}$. The flame will then show an increase in burning velocity governed roughly by the relation

$$\frac{S_t}{S_u} \approx \left(\frac{a + u'\bar{L}}{a} \right)^{1/2} \tag{3.37}$$

where S_t is the turbulent burning velocity and a is the thermal diffusivity.

Turbulence on such a small scale is rather uncommon in practical flames. Larger scale turbulence causes the flame front to distort and produce a *wrinkled flame*. The flame becomes thicker and more blurred in appearance giving rise to the term *flame brush*. Such a flame is more rounded in shape and quite noisy. The scale of the turbulence is now significantly greater than the flame front thickness and for some considerable time it was believed that transport processes were strictly molecular in the wrinkled flame. The increased rate of burning arises primarily from the increased area of the flame front and the turbulent flame consists essentially of an array of laminar flames. While this model of the wrinkled flame, primarily associated with the name of Damköhler [61], is basically correct, it represents only part of the

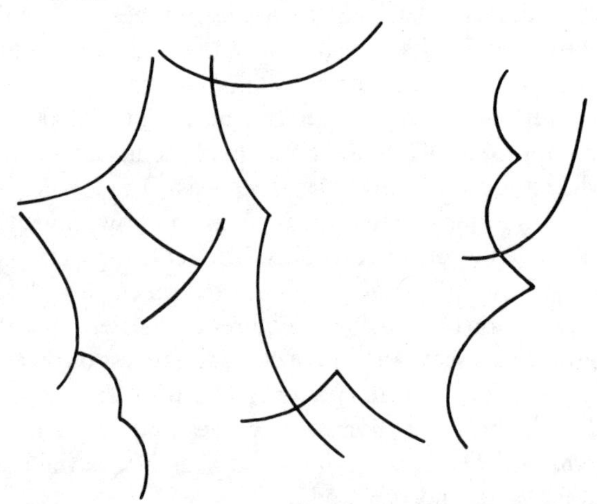

Fig. 3.14 Diagrammatic representation of cusp-shaped flame fronts in a turbulent flame (after Gaydon and Wolfhard [9]).

picture. To begin with, the front in the wrinkled flame is not sinusoidal in shape, but instead comprises a series of convex surfaces linked by cusps (Fig. 3.14). The three-dimensional structure in the vicinity of these cusps leads to an enhanced burning velocity. Even when averaged over the whole flame area, the effective velocity is well in excess of the anticipated laminar burning velocity.

This model neglects a significant feature of turbulence. As the turbulence intensity increases, vortex stretching of large eddies produces smaller and smaller eddies until ultimately the energy is dissipated in randomized molecular motion. Likewise, there is experimental evidence that combustion becomes increasingly concentrated within smaller length scales as turbulence increases. In the *two-eddy theory*, the complete turbulence spectrum is simplified into two principal eddy sizes: large eddies whose size is given by the integral length scale L and small dissipative eddies. Within both types of eddies, the rate of burning is expressed by the product of the rate of eddy decay and the amount of mixture reacted during the eddy lifetime. This theory, and others, suggest that the ratio S_t/S_u might be expressed as a function of u'/S_u for different values of Re_L and, at least for low values of u', the rather scanty and unreliable experimental data lends some support to this view [62, 63].

At even higher turbulence intensities, the flame front becomes much more distorted and pockets of burning gas may break away and travel in the hot burnt gas stream before being finally consumed. This leads to a thicker reaction zone and slower burning. Eventually flame stretch may extend to the point where holes appear in the front. Cold gas is then able to enter the reaction zone and reduce the temperature and reaction rate until the flame is eventually extinguished.

It is a limitation of theories of turbulent burning velocity that they do not predict this flame extinction with increasing turbulence. Recent work on laminar flames shows how straining the flow field can reduce the laminar burning velocity and lead to quenching. This effect might be expected to operate in turbulent flames and propagation would then only be possible when the ratio of chemical to eddy lifetime is less than some limiting value. These concepts suggest additional dependences of S_t/S_u upon both the Lewis number of the deficient reactant and a dimensionless activation energy for the reaction rate. The two-eddy theory can be modified to allow for partial or complete extinction of the flame at high strain rates, or short eddy lifetimes, leading to the result that S_t/S_u eventually flattens and even decreases slightly with increase in u'/S_u prior to final quenching [62, 63].

3.1.7 The well-stirred reactor

The ideas described above apply best to a burner intended to produce an ordinary flame. In some situations, for example, a jet engine, the burner and its container are designed to encourage recirculation which promotes high levels of turbulence. By these means, very high rates of heat release are achieved, and it is of some interest to explore the factors which control the heat release rate in such a system. The Longwell bomb [64] is an insulated reactor in which mixing is so efficient that the temperature and concentrations become uniform throughout. In such a reactor the term flame propagation loses its meaning and the combustion behaviour is determined by the characteristics of the chemical reaction and the flow rate of the fuel–oxidant mixture.

Fig. 3.15 shows an idealized Longwell bomb. The fuel–oxidant mixture, temperature T_0 and fuel concentration $[\mathrm{F}]_0$, enters at a volumetric flow rate v_0'. In the reactor the temperature is T and the fuel concentration [F]; the stream leaving the reactor at a flow rate v' has the same composition and temperature as the contents of the reactor itself.

The conservation equation for fuel can be written

$$v_0'[\mathrm{F}]_0 - V\boldsymbol{r} = v'[\mathrm{F}] \tag{3.38}$$

where V is the volume of the reactor and $\boldsymbol{r}$ is the rate of reaction per unit volume. If global kinetics which are first order in fuel concentration are assumed

$$\boldsymbol{r} = k[\mathrm{F}] \tag{3.39}$$

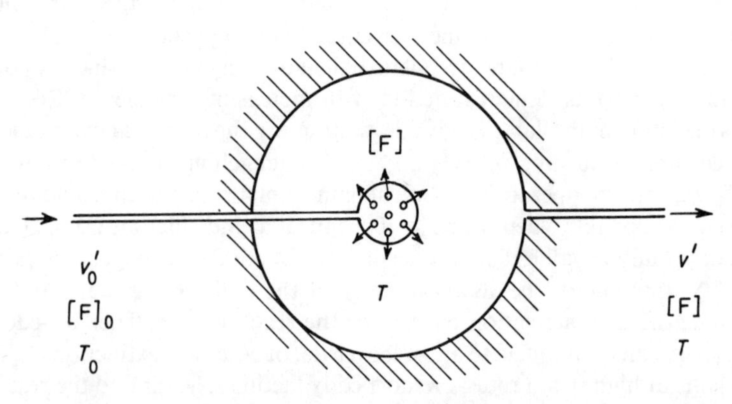

Fig. 3.15 The well-stirred reactor or Longwell bomb. v_0', flowrate; $[\mathrm{F}]_0$, fuel concentration; T_0, temperature.

and substituting Equation 3.39 in Equation 3.38 leads to the result

$$[\mathrm{F}] = \frac{v_0'[\mathrm{F}]_0}{(Vk + v')} \tag{3.40}$$

The rate of heat release is given by

$$\begin{aligned} \dot{q}_+ &= (-\Delta H)Vr \\ &= \frac{(-\Delta H)Vkv_0'[\mathrm{F}]_0}{(Vk + v')} \end{aligned} \tag{3.41}$$

Expressing k in the usual Arrhenius form and rearranging gives

$$\dot{q}_+ = \frac{(-\Delta H)v_0'[\mathrm{F}]_0 \exp(-E/RT)}{[\exp(-E/RT) + (v'/VA)]} \tag{3.42}$$

At low temperatures, the first term in the denominator is negligible compared with the second, and the rate of heat generation increases exponentially with temperature. At high temperatures, the first term in the denominator is much greater than the second and the heat generation rate approaches the value $(-\Delta H)v_0'[\mathrm{F}]_0$ corresponding to complete combustion of the fuel.

As the reactor is perfectly insulated, the only heat lost is that carried away by the stream of combustion products leaving the reactor and

$$\dot{q}_- = v'\rho c(T - T_0) \tag{3.43}$$

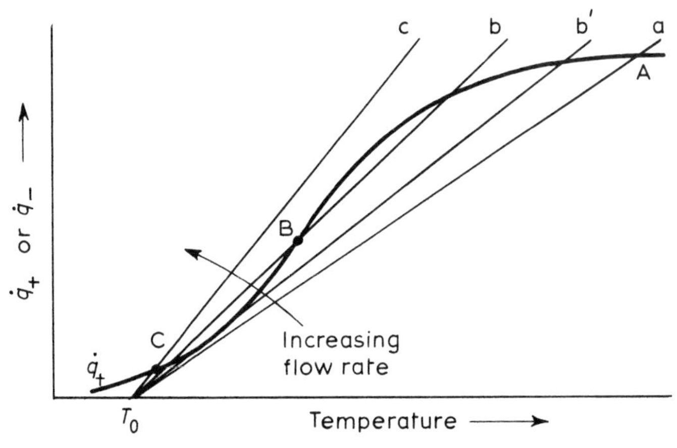

Fig. 3.16 Rate of heat generation, $\dot{q}_+$, and heat loss, $\dot{q}_-$, as a function of temperature in an adiabatic well-stirred reactor. The lines a, b, b′ and c refer to heat loss rates at increasing flow rates.

where ρ and c are the density and specific heat capacity, respectively, of the product stream. Since ρ and c do not change greatly with temperature, $\dot{q}_-$ depends principally on v' and T_0.

In Fig. 3.16 the S-shaped curve represents $\dot{q}_+$ as given by Equation 3.42 while the straight lines correspond to $\dot{q}_-$ with different values of v'. At low rates of heat loss (line a) the system will come to a steady-state at point A where heat loss and heat generation are equal. There is a large temperature rise from the initial value T_0, a high rate of reaction and virtually complete combustion: ignition has occurred spontaneously. If the flow rate is increased so that the heat loss rate is given by line c, then the steady state C is associated with a low temperature, slow reaction and only a small conversion of reactants.

This change from a high to a low reaction rate as the reactant flow rate is increased, described as *blow-out* or *flame extinction*, is of considerable practical importance.

With an intermediate flow rate (line b) there are three intersections; the upper and lower ones have the properties already described, while the central one B is an unstable situation in that any small displacement results in the system moving to either the upper or lower stable state. The critical condition for blow-out is when the heat loss line is tangential to the heat generation curve (line b′).

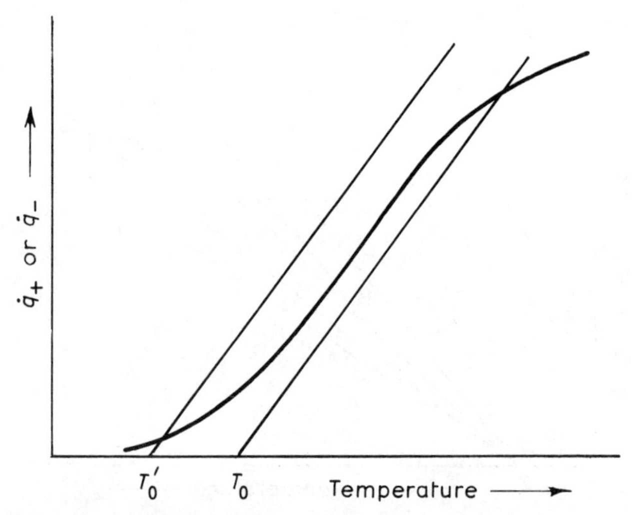

Fig. 3.17 Rate of heat generation, $\dot{q}_+$, and heat loss, $\dot{q}_-$, as a function of temperature in an adiabatic well-stirred reactor showing the effect of changing the temperature of the reactants at the inlet.

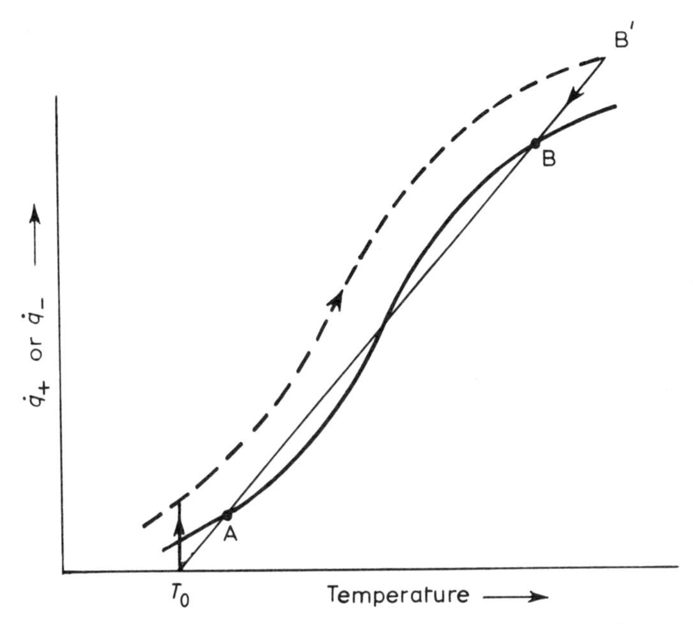

Fig. 3.18 Rate of heat generation, $\dot{q}_+$, and heat loss, $\dot{q}_-$, as a function of temperature in an adiabatic well-stirred reactor showing the effect of supplying energy from an external source for ignition.

Lowering the inlet temperature from T_0 to T'_0 while keeping the flow rate constant can also result in flame extinction (Fig. 3.17).

With an intermediate flow rate shown again in Fig. 3.18, cold reactants enter at T_0. Reaction will begin and the heat generation curve is above the heat loss line so the system will heat up until it settles down at the lower stable point A. The upper stable point B can only be reached if heat is temporarily added from an external source. Then the heat generation curve is raised to the dotted line and the system ignites going to the new upper stable point B′. When the external heat source is taken away, the system moves to the normal upper stable point B, where it will stay indefinitely.

Although this model is highly simplified, it is capable of extension to include more complex kinetics and other modes of heat loss.

3.1.8 Flame stabilization at high velocity

Typical laminar burning velocities for hydrocarbon–air mixtures are about 0.4 m s^{-1} and under turbulent flow conditions are unlikely to be much above 1.5 m s^{-1}. The Bunsen burner will accept flow velocities up to five times the burning velocity so that the maximum flow velocity which can be considered with this design of burner must be about $2\,m\,s^{-1}$.

The flow velocities in ram-jet and turbo-jet engines are about $50\,m\,s^{-1}$. Under these conditions, flow stabilization has to be achieved in a quite different manner. The principle adopted is to recirculate some of the hot combustion products so that they continually re-ignite the oncoming gas and thus prevent the flame from being blown away. This recirculation may be achieved either by re-directing some of the gas flow so that it travels normal to or even against the main flow (aerodynamic stabilization) or by inserting a suitable surface which produces eddies (bluff-body stabilization). Aerodynamic stabilization is normally associated with the small combustion chambers used in turbo-jet engines while bluff-body stabilization is more common in large combustors and is used in the after-burners of gas turbines and in ram-jet engines.

3.2 DIFFUSION FLAMES

3.2.1 Properties of diffusion flames

Diffusion flames, or jet flames, differ from pre-mixed flames in that combustion occurs at the interface between the fuel gas and the oxidant gas and the burning process depends more upon the rate of mixing than on the rates of the chemical processes involved. It is more difficult to give a general treatment of diffusion flames largely because no simple, measurable parameter, analogous to the burning velocity, can be used to characterize the burning process. Although less is therefore known about such flames they are nevertheless very important both academically and commercially.

Because the dominant physical process is that of mixing, it is possible to make a clear distinction between flames which involve the two different flow regions. In slow-burning diffusion flames, typified by candle flames, the fuel rises slowly and laminar flow ensues. The mixing process occurs solely by molecular diffusion and, as well as shown below, the properties of the flame are determined only by molecular quantities.

In industrial burners, and in gas turbines, where the fuel is usually introduced in the form of discrete droplets, burning is rapid, flow speeds are high and the mixing process is associated with the turbulence of the flow. The aerodynamics of the system will then dominate to a large extent the molecular properties.

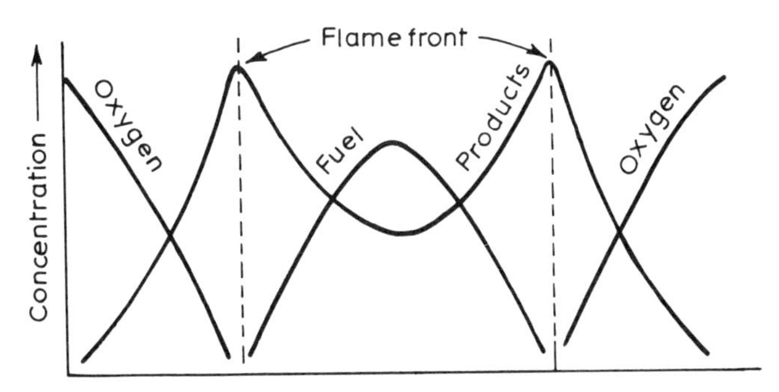

Fig. 3.19 Concentration profiles through a laminar diffusion flame.

As in the case of pre-mixed flames, various stability regions may be distinguished. The burner plays very little part in the stability compared with pre-mixed flames because the burning region is necessarily confined to the interface between the two gases and the burner serves simply as a nozzle to provide a directed stream of fuel. Flash-back is clearly impossible down to very low nozzle velocities and a very small flame can be stabilized on the nozzle. As the velocity increases, the flame maintains an essentially cylindrical shape and simply becomes taller. If a cross-section were taken through the flame (Fig. 3.19) it would be found that the fuel concentration has a maximum on the axis and falls away rapidly to the flame boundary. Similarly, the oxygen concentration decreases close to the flame and falls approximately to zero at the boundary. The concentration of products, on the other hand, has a maximum at the boundary, where the major extent of reaction is occurring, and falls away both towards the axis and into the surrounding atmosphere. The flame boundary strictly defines the surface at which combustion is complete but since reaction is normally very rapid it represents the position at which the fuel: oxygen ratio becomes stoichiometric. As the fuel concentration will decrease with height, the position of the boundary moves towards the axis and converges at the tip where all the fuel is consumed. In principle, therefore, the flame should be conical in shape, reaction being confined largely to a thin zone forming the flame boundary which corresponds closely to the reaction zone in the pre-mixed flame.

This model applies strictly only to the case where the fuel and oxygen are moving upwards at the same velocity. Although concentric burners which simulate this model are used, the commonest diffusion flame is that produced when a jet of fuel enters a stagnant atmosphere. The transfer of momentum close to the boundary leads to the formation of eddies. An eddy of combustion products causes the fuel and oxygen streams to become further separated so that diffusion becomes slower. The height of the flame then increases and the shape becomes more cylindrical. If the eddies break away, the fuel and oxygen are brought into more intimate contact so that the burning rate increases and the flame height is reduced. These eddy effects are responsible for the 'flicker' usually associated with diffusion flames. The flow is therefore only partially laminar although that near the axis may be little affected.

The character of the flame depends on the nature of the gas flow which is usually indicated by the Reynolds number. However, a rather interesting effect is observed with turbulent flow. The Reynolds number is inversely proportional to the kinematic viscosity, and the value of this parameter increases drastically in the combustion region. Thus turbulent flow in the burner, characterized by the high Reynolds number, may give rise to laminar flow at the flame boundary and hence the typical laminar flow behaviour with tall, cylindrical flames. Mixing is therefore occurring by molecular diffusion rather than by turbulent entrainment. The criterion must therefore refer to the type of flow in the region in which the flame boundary occurs.

If the nozzle velocity is increased, the laminar flame increases its

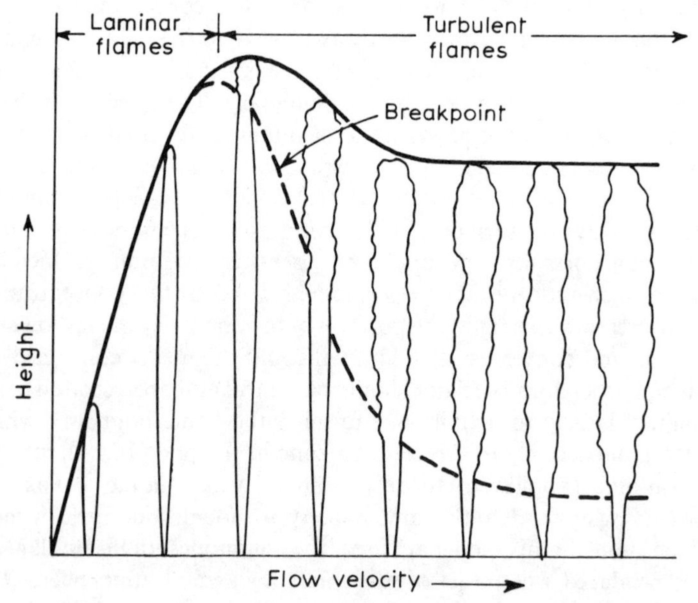

Fig. 3.20 Variation of shape of diffusion flame with flow velocity.

height almost linearly until turbulent mixing occurs. This appears first at the tip and moves down progressively with increase in velocity. The flame length also falls in this region, the visible boundary spreading outwards. Eventually a stage is reached at which both the height of the flame and the position of the transition, known as the *break-point* remain constant irrespective of any additional increase in nozzle velocity. These various effects are shown diagrammatically in Fig. 3.20.

In the turbulent flame region, the phenomenon of lift, similar to that observed with pre-mixed flames, may also occur. The reasons are not fully understood but the change in appearance suggests that entrainment of air at the burner rim in producing some of the characteristics of the pre-mixed flame and combustion occurs only when the fuel: oxygen ratio has fallen sufficiently. A velocity is eventually reached at which only lifted flames occur and beyond this region there will be a blow-out limit for the lifted flames. The nature of the flame in various flow and composition regions is illustrated in Fig. 3.21 [65] (cf. Fig. 3.8 for pre-mixed flames).

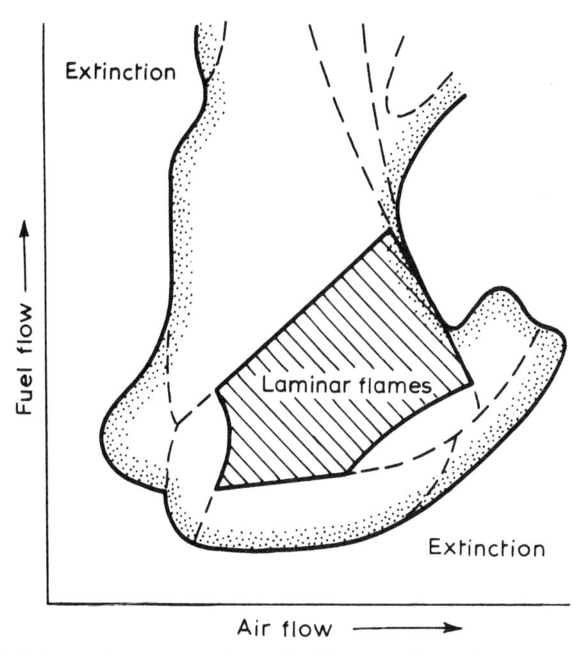

Fig. 3.21 Schematic representation of limiting conditions for various diffusion flames. The undesignated areas refer to meniscus flames, vortex flames etc. not described in the text (after Barr [65]).

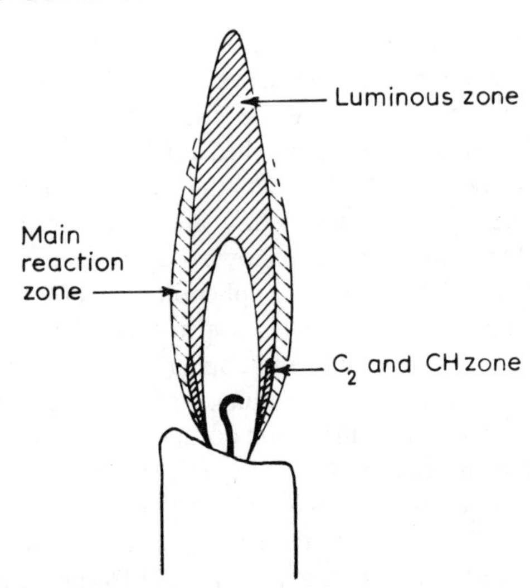

Fig. 3.22 Reaction zones in the candle flame (after Gaydon and Wolfhard [9]).

The model of a very narrow reaction zone is in fact highly idealized. Because of the transfer of heat as well as of matter, considerable reaction occurs on each side of the 'true' flame boundary. The general effect is for fuel to be pyrolysed and for the oxidant to form reactive radicals so that the reactants entering the 'true' reaction zone do not correspond to the initial gases. These reactions will be considered in more detail later but their existence is worth noting here because they can lead to the appearance of different reaction zones with different physical properties (Fig. 3.22) [9]. In hydrocarbon flames, the production of carbon particles is characterized by intense luminosity and the shape and behaviour of these luminous regions can be quite distinct from those of the 'normal' reaction zone.

3.2.2 Theory of diffusion flames

Diffusion flames are more difficult to deal with than pre-mixed flames because there is no simple parameter characteristic of the process, such as the burning velocity. The closest measurable property is the flame height and several treatments have been developed to relate observed flame heights to the properties of the gases.

The simplest analysis is due to Jost [66]. The Einstein diffusion equation states that the average square displacement $\overline{x^2}$ is given by

$$\overline{x^2} = 2Dt \tag{3.44}$$

The height of the flame is taken as the point where the average depth of penetration is equal to the tube radius r. Approximating $\overline{x^2}$ by r^2 this gives $t = r^2/2D$, but since the time t is also given by the height y divided by the gas velocity u (taken as constant), then

$$y = \frac{ur^2}{2D} \tag{3.45}$$

The volume flow rate v' is equal to $u\pi r^2$ so that

$$y = \frac{v'}{2\pi D} \tag{3.46}$$

Although very crude, this approximation permits certain predictions: for example, that the flame height at a constant volume flow rate should be independent of burner dimensions. Also, because the diffusion coefficient is inversely proportional to pressure, the height of the flame will be independent of pressure at a constant mass flow rate.

Some insight into the behaviour of turbulent diffusion flames is provided by this simple formula. The molecular diffusivity D may be replaced by the eddy diffusivity, which is the product of the scale of the turbulence L and a velocity fluctuation term. Since L is proportional to the tube diameter, this leads to the interesting conclusion that with a turbulent diffusion flame the height is proportional only to the orifice diameter.

The full mathematical analysis of the problem proves to be very difficult and a number of questionable approximations are necessary but the procedure may be outlined as follows. The reaction zone is defined as the region where the fuel and oxygen are in stoichiometric ratio; this definition is necessary but is clearly incorrect as reaction will be occurring over an extremely wide range of fuel:oxygen ratios particularly at high temperatures. The assumption is also made that the diffusion process is rate-determining so that the rate of reaction is related directly to the amounts of fuel and oxygen diffusing into the reaction zone. Although very satisfactory in principle this approximation is difficult to apply because the fundamental parameter required, the diffusion coefficient, varies with both temperature and composition. It is normal also only to consider diffusion in a radial direction [67].

Recent modifications to the original analysis have resulted in very satisfactory agreement being obtained between predicted and observed flame sizes on several types of burner which are used in domestic and industrial applications [68–70].

3.2.3 The chemistry of diffusion flames

Because little detailed knowledge is available on the chemistry of diffusion flames, it is preferable to review the situation here rather than in the section devoted to the chemistry of combustion processes.

Diffusion flames can be stabilized in various ways. The most usual is when a jet of fuel vapour is injected through a burner port or nozzle into an oxidizing atmosphere, but burners have also been designed in which two parallel streams, one of fuel and one of oxidant, flowing in adjacent ducts, are allowed to come into contact; such a burner gives a flat flame very suitable for spectroscopic examination. Another special burner giving a flat flame brings opposing flows of fuel and oxidant together giving a counterflow diffusion flame. Flames on wicks also provide examples of diffusion flames; in appearance these flames usually possess an inner blue zone and an outer yellow luminous zone. Samples taken from those zones have demonstrated that the primary chemical processes taking place in such diffusion flames involve

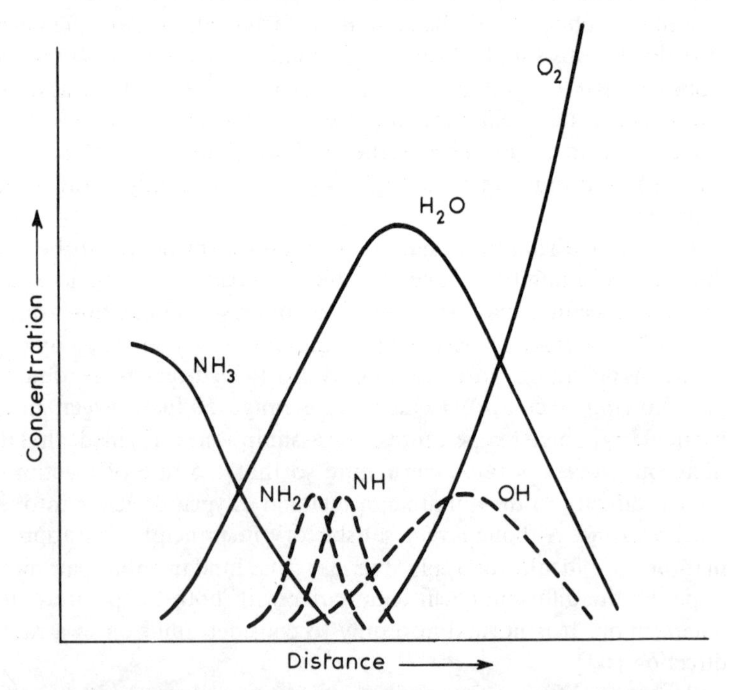

Fig. 3.23 Schematic representation of the concentration profiles in the ammonia–oxygen diffusion flame (after Wolfhard and Parker [74] and Dows *et al.* [75]).

pyrolysis of fuel and that oxidation of these pyrolysis products then takes place at the boundary of the inner flame zone [71–73].

In any diffusion flame, the fuel and oxidant do not really come into contact with each other. As Fig. 3.19 shows there is a boundary at which the concentrations of each fall essentially to zero. Reaction takes place largely by pyrolysis of the fuel due to diffusion of active species, notably OH radicals where oxygen is the oxidant, into the fuel. These reactions lead to intermediates derived from the fuel which diffuse into the oxidant zone. If the diffusion flame is flat, distinct zones, corresponding to high concentrations of various intermediates, may be observed. This has been illustrated for the candle flame (Fig. 3.22) and is further shown by the profiles for the ammonia–oxygen flame shown schematically in Fig. 3.23 [74, 75]. The progressive degradation of ammonia through NH_2 and NH on passing towards the oxygen zone is clearly seen. The dominant reactions involve NH_3 with OH on one side and NH_2 and NH with O_2 on the other. The water product spans both fuel and oxygen regions with a maximum approximately where both reactants fall to zero concentration.

An important feature of diffusion flames is that the radiation corresponds closely to that expected from local thermal equilibrium, indicating that the diffusion of active species into the reactants tends to reduce the probability of obtaining abnormally high local radical concentrations.

3.3 SUGGESTIONS FOR FURTHER READING.

Bradshaw, P. (1971) *An Introduction to Turbulence and its Measurement*, Pergamon, Oxford.

Fenimore, C.P. (1964) *Chemistry in Premixed Flames*, Pergamon, Oxford.

Fristrom, R.M. and Westenberg, A.A. (1965) *Flame Structure*, McGraw-Hill, New York.

Gaydon, A.G. and Wolfhard, H.G. (1979) *Flames, Their Structure, Radiation, and Temperature*, 4th edn, Chapman and Hall, London.

Karlovitz, B. (1956) Combustion waves in turbulent gases, in *Combustion Processes* (ed. B. Lewis, R.N. Pease, and H.S. Taylor), Oxford University Press, Oxford, p. 312.

Lewis, B. and Von Elbe, G. (1961) *Combustion, Flames and Explosions of Gases*, 2nd edn, Academic Press, New York.

Lewis, B. and Von Elbe, G. (1956) Combustion waves in non-turbulent explosive cases, in *Combustion Processes* (ed. B. Lewis, R.N. Pease and H.S. Taylor), Oxford University Press, Oxford, p. 216.

Wohl, K. and Shipman, C.W. (1956) Diffusion flames, in *Combustion Processes* (ed. B. Lewis, R.N. Pease and H.S. Taylor), Oxford University Press, Oxford, p. 365.

3.4 PROBLEMS

1. The value of the Peclet number for quenching in most hydrocarbon–air mixtures is about 50. Estimate the quenching distance in stoichiometric mixtures of air and (a) hydrogen, (b) methane, (c) propane and (d) acetylene at 298 K and a pressure of 1 atm given the following data:

	H_2	CH_4	C_3H_8	C_2H_2	Air
S_u (m s^{-1})	2.0	0.45	0.43	1.58	—
λ (W m^{-1} K^{-1})	0.151	0.0342	0.0266	0.0213	0.0261
C_p (J mol^{-1} K^{-1})	28.6	35.6	73.0	44.1	29.2

   What will be the corresponding values of the quenching diameter?
   (*Hint*: For the hydrogen–air mixture, the experimental value of λ is *ca.* 0.04 W m^{-1} K^{-1}; for the remaining properties of the fuel–air mixtures, use mean values based on the molar ratios.)
2. Using the information given in Question 1, calculate the quenching distance in a stoichiometric propane–air mixture at 298 K and 0.1 atm pressure.
3. The burning velocities of stoichiometric hydrocarbon–air mixtures are about 0.4 m s^{-1}. Using propane as a typical fuel, estimate the average reaction rate and the thickness of the pre-heating zone. The thermal conductivity of the reaction mixture may be taken as 0.025 W m^{-1} K^{-1} at 298 K and the mean flame temperature as 1000 K. (Use Table 1.2 to obtain values of c_p and assume that the heat capacity of air is the same as that of nitrogen).
4. Estimate the change in burning velocity of a stoichiometric propane–air mixture which would result if the nitrogen in air were replaced by helium, assuming that the average reaction rate and mean flame temperature are unchanged. At 298 K, the thermal conductivity of the reactants is 0.36 W m^{-1} K^{-1}.
5. Starting from Equation 3.7 prove that

$$\frac{[\text{Air}]}{[\text{Fuel}]} = \frac{[(1-\beta)^2 + 4\alpha\beta]^{1/2} - (\beta + 1)}{2}$$

   where $\alpha = A_B/A_F$ and $\beta = \rho_F/\rho_A$.
   Show that if $\beta \approx 1$ and α is large, this expression reduces to

$$\frac{[\text{Air}]}{[\text{Fuel}]} = (\alpha\beta)^{1/2}$$

6. How, in approximate numerical terms, should the fuel orifice diameter be altered so that a burner which currently operates on natural gas (methane) will burn (a) propane and (b) hydrogen with the same stability limits. Assume that the dependence of g_F and g_B on Φ remains unaltered.

4

Detonation waves in gases

The normal flame or combustion wave described in Chapter 3 travels at relatively low velocities, usually between 0.1 and 10 m s^{-1}. Most explosive mixtures can also display flame front velocities several orders of magnitude greater, for example, 2000–3000 m s^{-1}. These 'flames' exceed the characteristic sound speed of the gas and are termed *detonations* or *detonation waves*. Gaseous detonations are characterized by drastic changes in pressure and temperature: here again they differ from pre-mixed flames which display a marked temperature change but very little alteration in pressure. Clearly, the mechanism by which the flame propagates must differ in the two phenomena. In the simple flame, propagation depends on the transport of heat, and possibly also chemically reactive species, ahead of the flame front into the unburnt gas. Such transport processes are too slow to explain detonation, and reaction in this case is initiated by a supersonic pressure, or *shock*, wave. The shock wave heats the reactive gas extremely rapidly and the energy liberated by the subsequent chemical reactions provides the driving force for the shock. Before discussing detonations in greater detail it is useful to summarize the principal features of a simple shock wave.

4.1 SHOCK WAVES

Shock waves are associated with a large number of physical phenomena, both natural and man-made. The commonest examples are provided by aircraft flying at supersonic speeds and by explosions or volcanic eruptions. The shock wave is characterized by a virtually instantaneous pressure rise associated with the rapid release of a substantial amount of energy.* If this energy release is not maintained,

* The term 'wave' is hardly descriptive of such an abrupt change, and some authors prefer to omit it and refer simply to a shock.

the shock will be accompanied by a more gradual expansion, *or rarefaction*, wave which returns the pressure to the ambient value or below.

A shock wave can occur in any fluid medium and may be defined as a step transition or discontinuity in the physical properties of the medium propagating without change. A sound wave progresses as a sequence of pressure pulses and, provided the pulses are of low amplitude, the velocity has a constant value characteristic of the medium. If the pulses are of greater amplitude, the velocity increases by an amount which depends on the amplitude and the wave becomes supersonic. A disturbance which is originally sinusoidal in shape will become progressively more distorted, the region of increasing pressure becoming steeper and that of falling pressure becoming shallower (Fig. 4.1).

An explanation of the way in which a shock wave forms is provided by the following simple model. The steady acceleration of a piston in a cylindrical tube is regarded as being composed of a series of

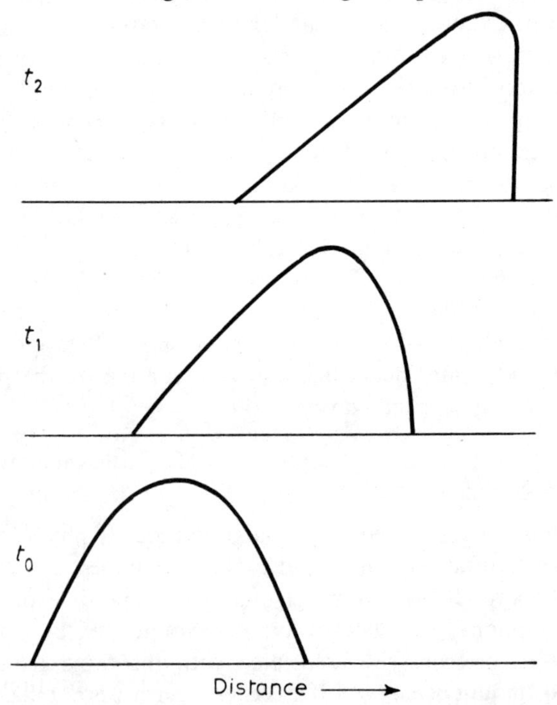

Fig. 4.1 Development of a finite sinusoidal pressure pulse into a shock wave; t_0, t_1 and t_2 refer to successive times.

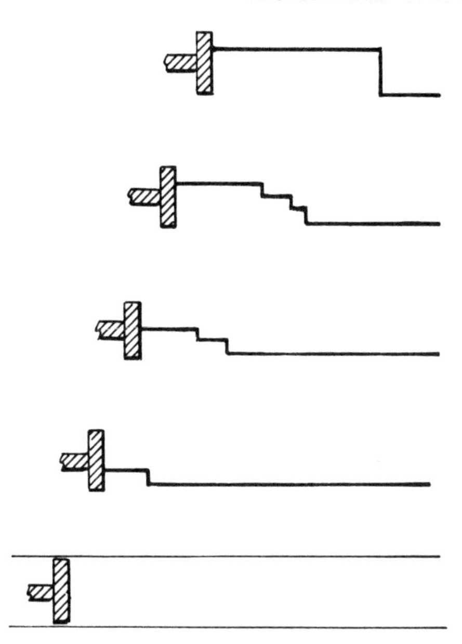

Fig. 4.2 Formation of shock wave from the acceleration of a piston in a tube.

infinitesimal accelerations (Fig. 4.2). Each of these generates in the gas ahead of the piston a pressure pulse which travels at the appropriate sound speed. In addition to imparting movement to the gas, each pulse heats the gas adiabatically thereby raising the sound speed in the gas. The following impulse therefore travels faster than the one which preceded it and must catch it up. Eventually, these very small pulses coalesce to form a pressure pulse of finite amplitude travelling at a velocity greater than that of sound in the undisturbed gas. Shock waves may be generated in the laboratory in a manner which resembles that just described.

It should be noted that the gas through which the shock wave has passed will have a higher pressure, density and temperature than the undisturbed gas and will be moving in the same direction as the shock front itself, although with a lower velocity. Also, to preserve the discontinuity a continuous source of energy is required behind the front. In the detonation wave, this energy is provided by chemical reaction in the heated gas.

The properties of the gas on each side of the transition can be examined without a knowledge of the processes occurring in the transition itself since conditions are essentially uniform before and after

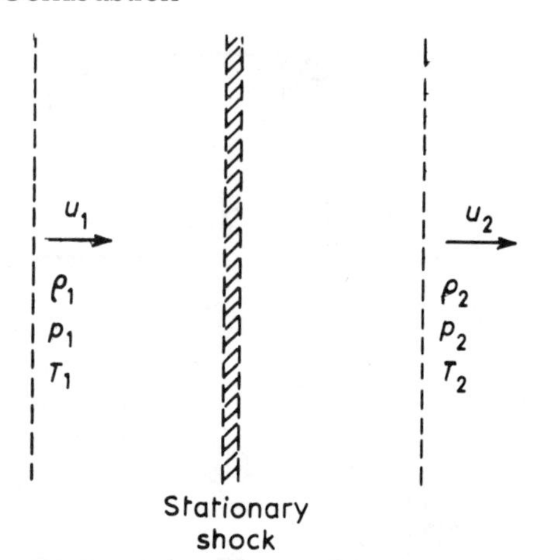

Fig. 4.3 Notation used in the description of a stationary shock wave.

passage of the shock and the whole process is considered to be one-dimensional. The system is most conveniently treated by referring it to a co-ordinate system moving with the front (Fig. 4.3). Application of the conservation equations for mass and momentum leads to the following expressions:

Conservation of mass (continuity)

$$\rho_1 u_1 = \rho_2 u_2 \quad \text{or} \quad u_1/v_1 = u_2/v_2 \tag{4.1}$$

Conservation of momentum

$$p_1 + \rho_1 u_1^2 = p_2 + \rho_2 u_2^2 \quad \text{or} \quad p_1 + u_1^2/v_1 = p_2 + u_2^2/v_2 \tag{4.2}$$

where v, the specific volume (volume per unit mass) $= 1/\rho$.

It should be noted that these equations take no direct account of the physical state or chemical nature of the medium.

Manipulation of these two expressions leads to the mechanical shock relationships

$$u_1 = v_1 \left(\frac{p_2 - p_1}{v_1 - v_2} \right)^{1/2} \tag{4.3}$$

$$u_2 = v_2 \left(\frac{p_2 - p_1}{v_1 - v_2} \right)^{1/2} \tag{4.4}$$

The shock velocity in stationary co-ordinates, U_s, is equal to $-u_1$, so that we can obtain expressions for the *shock velocity*

$$U_s = v_1 \left(\frac{p_2 - p_1}{v_1 - v_2} \right)^{1/2} \tag{4.3a}$$

and for the *particle*, or *flow*, *velocity*

$$W = (v_1 - v_2) \left(\frac{p_2 - p_1}{v_1 - v_2} \right)^{1/2} \tag{4.5}$$

(note that the direction of motion has been inverted to avoid carrying an unnecessary minus sign).

It is often convenient to rewrite these velocities in dimensionless form by dividing by the sound speed in the undisturbed gas in which case the resulting quantities are termed *Mach numbers* and represented by the symbol *Ma*. The sound speed, a, is defined [6] by

$$a^2 = \left(\frac{\partial p}{\partial \rho} \right)_s = -v^2 \left(\frac{\partial p}{\partial v} \right)_s \tag{4.6}$$

and for an ideal gas it follows that

$$a = (\gamma p v)^{1/2} = \left(\frac{\gamma R T}{M} \right)^{1/2} \tag{4.7}$$

where γ is the specific heat ratio and M is the molar mass. It will be seen from Equations 4.3a and 4.6 that in the limit as p_2 approaches p_1, the shock velocity falls to the sound velocity a_1.

In order to proceed further we have to make use of the third conservation equation

Conservation of energy

$$e_1 + \frac{u_1^2}{2} + \frac{p_1}{\rho_1} = e_2 + \frac{u_2^2}{2} + \frac{p_2}{\rho_2}$$

or

$$e_1 + \frac{u_1^2}{2} + p_1 v_1 = e_2 + \frac{u_2^2}{2} + p_2 v_2 \tag{4.8}$$

Here e denotes specific internal energy, that is, energy per unit mass, $u^2/2$ flow energy and p/ρ pressure energy. The three conservation equations can be combined to give the *Rankine–Hugoniot* relationship

$$e_2 - e_1 = \tfrac{1}{2}(p_2 + p_1)\left(\frac{1}{\rho_1} - \frac{1}{\rho_2} \right) = \tfrac{1}{2}(p_2 + p_1)(v_1 - v_2) \tag{4.9}$$

Alternatively we may introduce the specific enthalpy, h, equal to $e + p/\rho$, giving the relationship in the form

$$h_2 - h_1 = \tfrac{1}{2}(p_2 - p_1)(v_1 + v_2) \tag{4.10}$$

It is important to note that energy and enthalpy are properties which depend on the chemical nature of the system so the precise form of these equations will vary from one medium to another. Since e and ρ are themselves functions of p and T, a complete solution for a particular medium requires a knowledge of the appropriate equations of state, i.e.

$$e = f_1(p, T) \tag{4.11}$$

$$\rho = f_2(p, T) \tag{4.12}$$

For the time being, we may assume that this information is available in either numerical or analytical form.

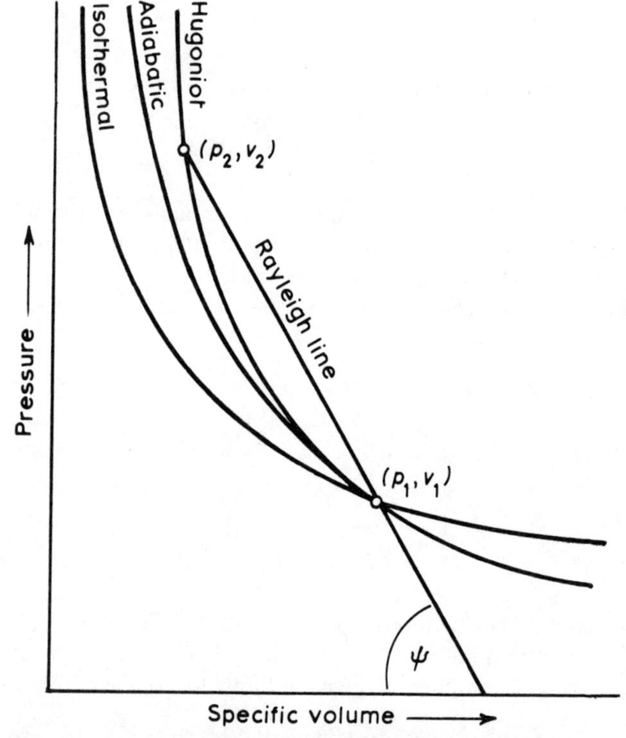

Fig. 4.4 Loci of points attainable by shock (Hugoniot), adiabatic, and isothermal transitions. The slope of the Rayleigh line connecting the points describing the initial and final states gives a measure of the velocity.

Equations 4.9 and 4.10 describe a curve on a p–v plot which connects states that can be attained by a shock transition. The curve, which is known as the *Hugoniot*, has a similar significance to the curves which relate states which may be achieved by an isothermal or adiabatic process (Fig. 4.4).

The chord joining the initial (p_1, v_1) and final (p_2, v_2) states is termed the *Rayleigh line*: if this line makes an angle ψ with the abscissa (Fig. 4.4), Equation 4.3a for the shock velocity can be rewritten

$$U_s = v_1(\tan \psi)^{1/2} \tag{4.13}$$

It has already been seen that in the weak shock limit the shock velocity reduces to the sound velocity: at this point, the Rayleigh line, the tangent, the adiabatic curve and the Hugoniot curve, all coincide. We should also note from the definition of the sound speed given in Equation 4.6 that the sound speed in the shock-heated gas is related to the tangent at (p_2, v_2) by

$$a_2 = v_2\left(\frac{-\partial p}{\partial v}\right)_s^{1/2} \tag{4.14}$$

Throughout this discussion, it has been assumed that the transition is very sharp although since any medium is composed of discrete molecules the shock front thickness must be finite. Measurements show that it is indeed very thin, corresponding to a few mean free paths in the gas ahead. At atmospheric pressure this means about 10^{-7} m, which is equivalent to 10^{-9} s.

Although the discussion in this chapter is directed mainly towards

Table 4.1 Typical shock wave properties in air and water (after Penney and Pike [76].)

Medium	Sound speed ($\mathrm{m\,s^{-1}}$)	Shock Mach number	Flow Mach number	Temperature rise (K)	Pressure ratio
Air	347	4.12	3.29	900	20
		9.03	7.87	3440	100
		12.6	11.3	5090	200
		19.7	18.1	7940	500
Water	1532	1.15	0.07	5	2000
		1.52	0.28	36	10000
		1.88	0.45	70	20000
		2.62	0.81	152	50000

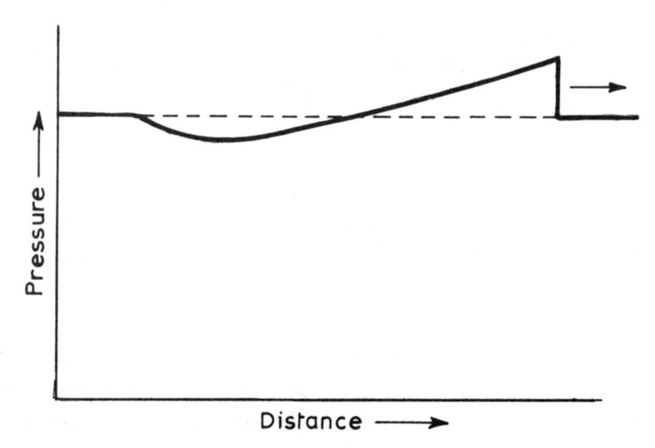

Fig. 4.5 Pressure profile of a blast wave.

gaseous media, the same analysis applies to condensed phases. In such phases, the compressibility is much lower and the sound speed defined by Equation 4.6 is therefore much greater. The relevant equations of state also reflect these differences in compressibility. Some typical shock parameters for air and water are listed in Table 4.1.

Shock waves are generated by explosions. Since the source is a single event, energy is not provided continuously and the shock front in the surrounding atmosphere is immediately followed by an *expansion* or *rarefaction* wave which 'eats away' at the front. The pressure profile therefore changes shape as the wave moves farther from the source: a schematic representation of the profile at an instant in time is shown in Fig. 4.5. In such cases the complete pressure profile is sometimes referred to as a shock wave, although in order to avoid confusion the authors prefer the term *blast wave*.

4.2 ONE-DIMENSIONAL STRUCTURE OF DETONATION WAVES

The shock wave behaves as a strictly one-dimensional phenomenon and in this chapter we shall apply the same restriction to detonations although, as we shall see later, this is not entirely justified. In writing the equation of state $e = f(p, T)$ for the gas, it was assumed that no chemical reaction had taken place. For an exothermic reaction system, in determining the final state (p_2, v_2) an allowance must be made for the energy liberated in the reaction. The effect of this is to displace the Hugoniot curve to higher values of p and v so that it no longer passes

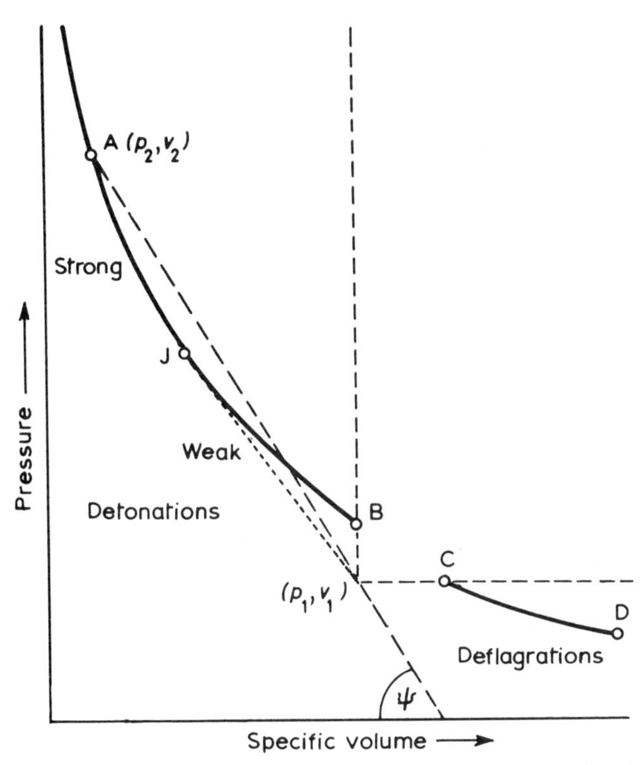

Fig. 4.6 Hugoniot curve for transition involving exothermic chemical reaction.

through (p_1, v_1) (Fig. 4.6). The Hugoniot still represents the solution to Equations 4.9 or 4.10 and now defines the states which are mathematically possible from a given initial state by a shock transition accompanied by the release of energy from a chemical reaction. Not all these states are physically real, however. Between the points B and C (Fig. 4.6), $(p_2 - p_1)/(v_1 - v_2)$ is negative so that the value of U_s given by Equation 4.3a is imaginary: a transition from (p_1, v_1) to any point between B and C therefore cannot correspond to a realizable situation. There are thus two quite separate regions: AB, which corresponds to $p_2 > p_1$, $v_2 < v_1$, and CD corresponding to $p_2 < p_1, v_2 > v_1$.

In the region CD to the right of (p_1, v_1), the slope of the Rayleigh line is less than the angle made by tangent at (p_2, v_2), so that the combustion is subsonic. Furthermore, the pressure and density decrease across the transition and the gas leaves it at a greater velocity than that at which it enters (this follows from Equations 4.3 and 4.4). These properties are precisely those described in Chapter 3 and this portion of the curve corresponds to a combustion wave or *deflagration*.

Points on the curve AB to the left of (p_1, v_1) correspond to velocities greater than the sound speed in the ambient gas and hence this portion of the curve describes the *detonation* process. The Rayleigh line drawn through (p_1, v_1) which makes a tangent to the curve corresponds to the minimum attainable detonation velocity: all other Rayleigh lines through (p_1, v_1) have a steeper slope and correspond to higher velocities. Those lines also indicate two possible final states. The detonation with the minimum velocity is termed a Chapman–Jouguet detonation and the final state described by point J to which this detonation refers is the Chapman–Jouguet or *C–J state* [77–80]. Detonations associated with final states to the left of the C–J state in Fig. 4.6 are referred to as *strong* detonations while those corresponding to points to the right are termed *weak* detonations.

On the strong detonation section of curve AB, the tangent at (p_2, v_2) has a steeper slope than the corresponding Rayleigh line. The front is then subsonic with respect to the gas behind it and any perturbations generated behind the front will catch up with it and hence cause the detonation to be attenuated. Strong or *over-driven* detonations may occur, for example, when initiated by a very strong shock wave, but unless they are supported by energy additional to that of the reaction they are unstable and decay to C–J detonations.

On the weak detonation section of curve AB, the detonation is supersonic with respect to the following gas. By an argument similar to that above, this means that the whole of the reaction energy cannot contribute to the motion of the front. Thus, it may be argued that while all final states on AB are allowed by the conservation equations only the C–J condition corresponds to a stable self-supporting detonation. This conclusion can also be reached by other more rigorous arguments.

Since the Rayleigh line to the C–J point is tangential to the Hugoniot, the sound speed a_2 in the burnt gas is given by Equation 4.14 in the form

$$a_2 = v_2\left(\frac{p_2 - p_1}{v_1 - v_2}\right)^{1/2} \tag{4.15}$$

Combining this with Equations 4.3a and 4.5 leads to the relationship

$$\boldsymbol{D} = W + a_2 \tag{4.16}$$

The detonation velocity $\boldsymbol{D}$ is thus equal to the sum of the particle, or flow, velocity and the sound speed in the burnt gas. This may be

Table 4.2 Typical detonation wave properties in gaseous mixtures. The detonability limits (after Lewis and Von Elbe [32]) apply to atmospheric pressure: the other figures, for stoichiometric mixtures at atmospheric pressure, are from a variety of sources using different methods and should only be considered as approximate.

Reactants	Detonability limits (% by volume)		Detonation velocity ($m\,s^{-1}$)	Detonation temperature (K)	Detonation pressure (atm)
	Lower	Upper			
$H_2 + O_2$	15	90	2825	3700	18.05
$CO + O_2$	38	90	1760	3500	18.6
$CH_4 + O_2$			2322	3700	
$C_2H_2 + O_2$	3.5	92	2350	4200	44
$NH_3 + O_2$	25.4	75	2400		
$C_3H_8 + O_2$	3.1	37	2350		
H_2 + air	18.2	59	1940	2950	15.6
C_2H_2 + air	4.2	50	1900	3100	19
CH_4 + air			1800	2740	17.2
C_3H_8 + air			1800	2820	18.3

regarded as a mathematical statement of the *Chapman–Jouguet postulate*. In physical terms it means that the detonation front moves at the sound velocity with respect to the gas behind it.

The importance of the C–J postulate is that it enables the detonation velocity, and hence the associated properties, to be predicted from the thermodynamic properties of the burnt gas. In principle, a detonation should be possible for all exothermic reactants but, in practice, kinetic factors are involved and composition limits are imposed on detonable mixtures. Some properties of typical gaseous detonations are given in Table 4.2.

This picture of the detonation wave is oversimplified because it assumes that chemical reaction proceeds to completion instantaneously. A better representation suggested independently by Zeldovich [81], von Neumann [82] and Döring [83] is to consider a family of Hugoniot curves corresponding to successive fractions of reaction (Fig. 4.7). As the detonation is a steady phenomenon, that is, one in which the detailed behaviour remains constant when referred to a co-ordinate system moving at a uniform velocity, all states through which the system passes must lie on the single Rayleigh line corresponding to the detonation velocity. The initial shock transition is to a point A on the non-reactive Hugoniot. The state of the gas then changes along the Rayleigh line in the direction of progressively greater

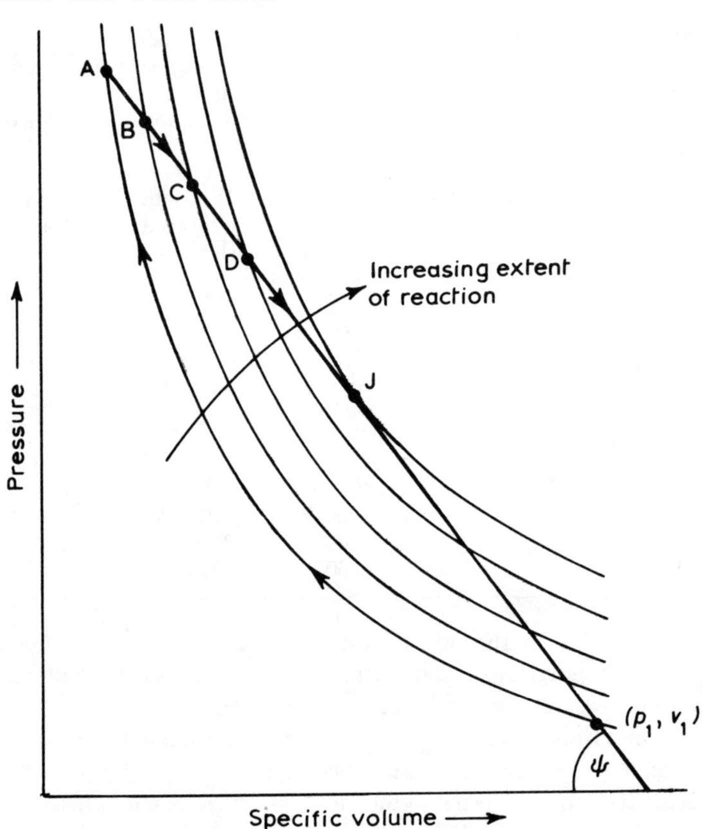

Fig. 4.7 Representation of a finite reaction zone by a family of Hugoniot curves corresponding to different extents of reaction.

extents of reaction until it reaches the C–J point (J) on the fully reacted Hugoniot.

In this *ZND model*, the detonation is thus visualized as a simple shock transition followed by a reaction zone of finite length within which the pressure and density fall and the temperature rises. The reaction zone is expected to have a width of about a millimetre or less, corresponding to about 0.5 μs, at atmospheric pressure. It is very difficult to measure reaction zone thicknesses but indirect methods have suggested reaction times of 0.3 to 0.5 μs in benzene–oxygen mixtures [84]. The model also predicts the existence of a sharp maximum in the pressure and density profiles (Fig. 4.8), referred to as the *von Neumann spike*, which has been observed experimentally

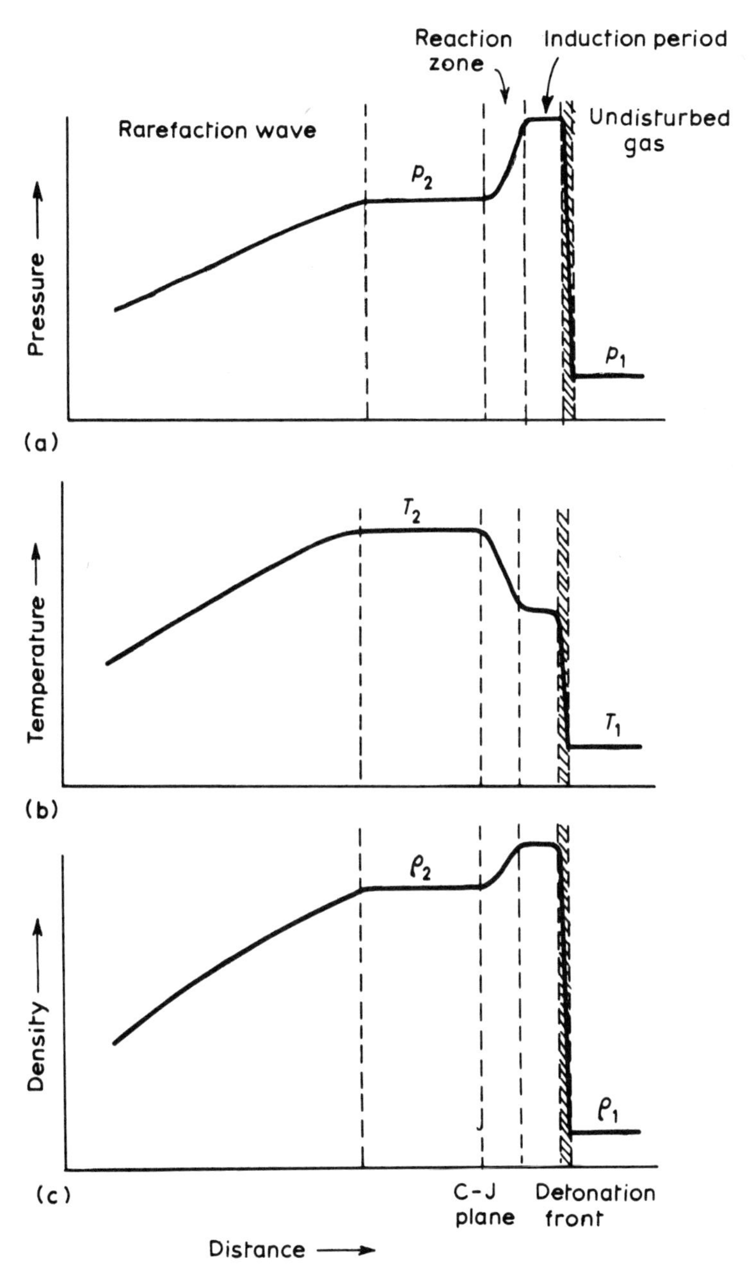

Fig. 4.8 Variation of (a) pressure, (b) temperature, and (c) density, through a detonation.

[85, 86]. The whole region between the front and the C–J plane represents a steady flow process and propagates without change.

The ZND model also provides further support for the C–J postulate. While final states characteristic of strong detonations are accessible by the mechanism described above and indeed are known to occur, if only as transients, the states characteristic of weak detonations cannot be achieved.

The burnt gas which leaves the C–J plane is moving in the same direction as the detonation but at the particle velocity and, in the absence of an additional source of energy, it must eventually come to a standstill. The mechanism is best understood by considering what happens when the detonation is initiated at the closed end of a long tube. At this end the gas is forced to remain motionless and so an expansion wave must follow the detonation front, cool the burnt gas, and accelerate it in the reverse direction. An expansion wave is necessarily isentropic and therefore travels at a velocity equal to the sound velocity of the gas superimposed upon the particle velocity. This is precisely the condition at the C–J plane to which the head of the expansion wave is therefore 'tied'. The rarefaction region is non-steady since the profiles there change with time as the detonation propagates through the medium. The behaviour is closely similar to that described earlier for a blast wave. It will be appreciated that steady motion of the detonation front is only possible if the flow behaviour of the burnt gas matches that of the expansion head thus providing an additional argument in support of the C–J condition for self-sustaining detonation.

4.3 MATHEMATICAL TREATMENT OF DETONATION

A complete solution of the Rankine–Hugoniot equations to give the C–J detonation velocity is much simpler than the corresponding calculation of the burning velocity of a pre-mixed flame because it depends only on the initial and final states of the gas. Even so, an analytical solution is not feasible unless a number of drastic assumptions are made. As far as possible, the approximations introduced below parallel those used in the elementary treatment of flame propagation developed in Chapter 3.

We begin by assuming that the fluid behaves as ideal gas with a constant specific heat capacity and a fixed molecular weight. Chemical reaction is accompanied by a heat release q per unit mass of gas. The following thermodynamic relations then apply

$$\frac{p_1 v_1}{T_1} = \frac{p_2 v_2}{T_2} = R' \tag{4.17}$$

$$h_1 = c_p T_1 \qquad h_2 = c_p T_2 \tag{4.18}$$

$$\gamma = c_p / c_V \quad c_p - c_V = R' \tag{4.19}$$

Note particularly that the enthalpies, specific heat capacities, specific volumes and the gas constant R', like the heat release, are referred throughout to unit mass of reacting gas. The Rankine–Hugoniot relation, Equation 4.10, may then be written

$$\frac{\gamma}{\gamma - 1}(p_2 v_2 - p_1 v_1) - q = \tfrac{1}{2}(p_2 - p_1)(v_2 + v_1) \tag{4.20}$$

A more convenient form of this equation is obtained by making the following substitutions

$$\pi = p_2/p_1; \quad \Gamma = v_2/v_1; \quad B = 2(\gamma - 1)q/p_1 v_1 \tag{4.21}$$

to give

$$2\gamma(\pi\Gamma - 1) - B = (\gamma - 1)(\pi - 1)(\Gamma + 1) \tag{4.22}$$

The (negative) slope of the tangent to the Hugoniot curve with co-ordinates π, Γ is obtained by differentiating Equation 4.22 to give

$$-\frac{d\pi}{d\Gamma} = \frac{(\gamma + 1)\pi + (\gamma - 1)}{(\gamma + 1)\Gamma - (\gamma - 1)} \tag{4.23}$$

The C–J condition is satisfied by equating this derivative to the (negative) slope of the Rayleigh line, that is, $(\pi - 1)(1 - \Gamma)$ in these generalized co-ordinates. From this equality, one obtains the relation

$$\pi = \frac{\Gamma}{(\gamma + 1)\Gamma - \gamma} \tag{4.24}$$

It proves particularly convenient to employ as a parameter the detonation velocity expressed as the dimensionless Mach number

$$Ma = \frac{D}{a_1} = \frac{D}{(\gamma p_1 v_1)^{1/2}} \tag{4.25}$$

From the shock relations above (Equation 4.3a)

$$Ma^2 = \frac{1}{\gamma}\frac{(\pi - 1)}{(1 - \Gamma)} \tag{4.26}$$

Substituting Equation 4.26 in Equation 4.24 and rearranging gives

$$\Gamma = \frac{1}{(\gamma+1)}\left(\frac{1}{Ma^2}+\gamma\right) \tag{4.27}$$

Introducing this expression into Equation 4.24 yields

$$\pi = \frac{Ma^2}{(\gamma+1)}\left(\frac{1}{Ma^2}+\gamma\right) \tag{4.28}$$

The expressions for π and for Γ can now be substituted into the Hugoniot Equation 4.22 to give the quadratic

$$(Ma^2)^2 - CMa^2 + 1 = 0 \tag{4.29}$$

where

$$C = [B(\gamma+1)/\gamma] + 2$$

The treatment so far applies equally to the detonation branch and the deflagration branch of the curve in Fig. 4.6 but we are concerned here only with the higher root, that is,

$$Ma^2 = \frac{C+(C^2-4)^{1/2}}{2} = \frac{C+C(1-4/C^2)^{1/2}}{2} \tag{4.30}$$

The term in the brackets may be expanded using the series

$$(1+x)^n = 1 + nx + n(n-1)x^2/2 + \cdots \tag{4.31}$$

If $C \gg 1$, as is usually the case, only the first term need be considered so that

$$Ma^2 \approx (\gamma+1)B/\gamma = 2q(\gamma^2-1)/\gamma p_1 v_1 \tag{4.32}$$

and

$$\boldsymbol{D} = [2q(\gamma^2-1)]^{1/2} \tag{4.33}$$

Equation 4.32 may be combined with Equations 4.27 and 4.28 to give

$$\Gamma = \gamma/(\gamma+1) \tag{4.34}$$

$$\pi = \gamma Ma^2/(\gamma+1) \tag{4.35}$$

The particle velocity follows from Equation 4.5, that is,

$$W(1-\Gamma)\boldsymbol{D} = [2q(\gamma-1)/(\gamma+1)]^{1/2} \tag{4.36}$$

From Equations 4.34 and 4.35

$$T_2/T_1 = \gamma^2 Ma^2/(\gamma+1)^2 \tag{4.37}$$

An alternative expression, obtained by substituting for Ma^2, is

$$T_2 = 2q\gamma/c_V(\gamma + 1) \tag{4.38}$$

It is interesting to note that, at the same level of approximation, the adiabatic constant volume temperature rise, $T_V = q/c_V$, so that

$$T_2/T_V = 2\gamma/(\gamma + 1) \tag{4.39}$$

In view of the number of approximations made, the calculated detonation properties for fuel–air mixtures turn out to be surprisingly close to those listed in Table 4.2. For fuel–oxygen mixtures, this simplified treatment yields less satisfactory results, but when allowance is made for dissociation of products at the higher temperatures involved, the agreement between the calculated and experimental detonation parameters is greatly improved [87].

4.4 THREE-DIMENSIONAL STRUCTURE OF DETONATIONS

Experimental measurements of detonation velocities agree well with the predictions based on the Chapman–Jouguet theory, typically to within $\pm 2\%$, except for small diameter tubes or close to the detonability limits when deviations of 10% or more are observed. In such cases, three-dimensional effects can be seen, the most notable being *spinning detonation* in which the front is tilted and rotates about the axis as it travels down the tube.

For many years it was believed that such three-dimensional effects were associated only with marginal situations. However, it is now known that the presence of exothermic reaction kinetics in the reaction zone renders the one-dimensional system unstable. If longitudinal instability occurs then *galloping detonations* are observed in which the detonation wave repeatedly decays and then jumps back to its original velocity. All stable detonations appear to show complex, three-dimensional structure.

One way of visualizing the origin of instability is as follows. The detonation front is represented by a square wave model in which the non-reactive shock is followed by a chemical induction period at the end of which the chemical energy is released instantaneously (Fig. 4.9(a)). In view of the exponential build-up of reactions in explosive systems (Chapter 2) such a model is quite realistic. Let us now examine what happens if the ignition front is given a minor perturbation so that it becomes distorted in a direction perpendicular to the flow (Fig. 4.9(b)).

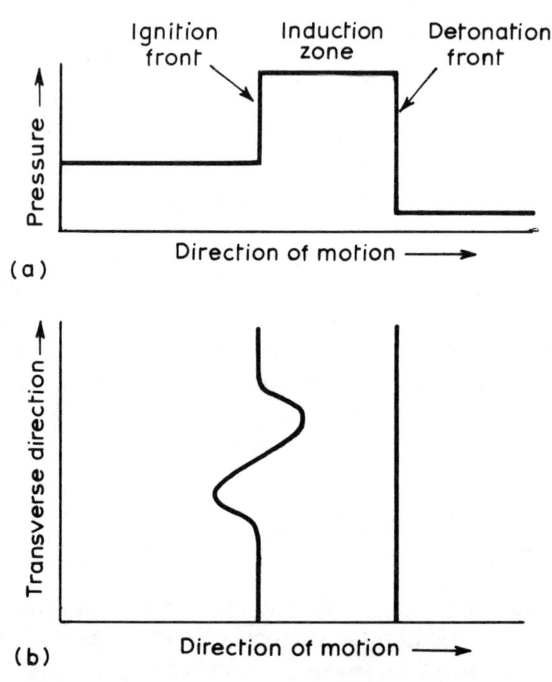

Fig. 4.9 (a) Model showing origin of transverse instabilities in a detonation wave. (b) Properties perpendicular to the front (after Shchelkin and Troshin [88]).

Small regions of high and low pressure now exist side-by-side so that the high-pressure region expands sideways to damp out the disequilibrium. In a non-reactive situation that would be the end of the matter. In the detonation, however, the expansion of the gas causes it to cool isentropically and the ignition delay (or induction period) being temperature dependent, increases significantly. Thus, instead of decaying the perturbation is amplified. It turns out that such an analysis actually underestimates the instability of detonations. It serves to show that three-dimensional effects are likely to be greater for reactions with higher activation energies and longer reaction zones which tend to be associated with marginal conditions.

In order to understand the three-dimensional structure of a steady detonation, it is necessary first to consider what happens when a shock wave strikes a rigid surface. As the gas adjacent to the surface must be brought to a standstill with respect to motion normal to the wall, a backward facing shock is reflected into the oncoming gas behind the incident shock (Fig. 4.10(a)). If the surface is inclined at an angle to the flow, the wave pattern in Fig. 4.10(b) is established. The point of intersection of the incident and reflected shocks travels upwards along the wall, the velocity and angle of the latter being such as to satisfy the constraint regarding motion normal to the wall. This is termed *regular reflection.*

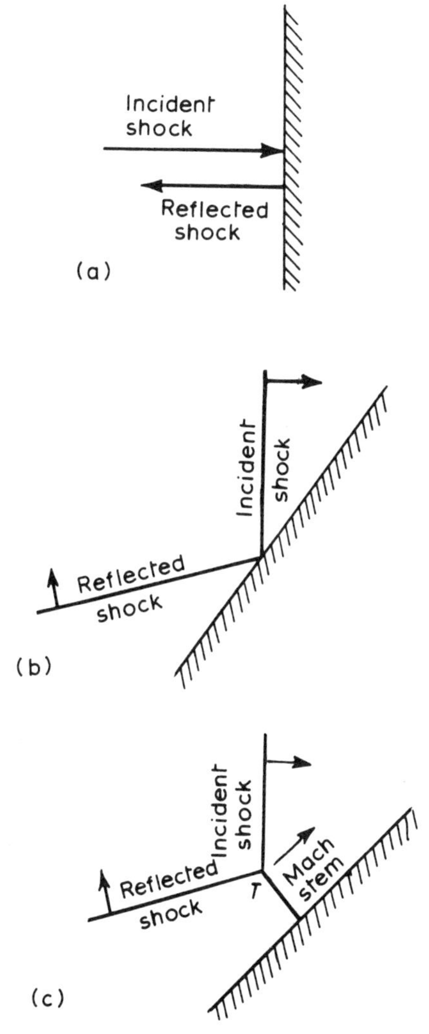

Fig. 4.10 Shock reflection (a) normal to a plane, (b) at an angle to a plane, (c) formation of a Mach stem.

As the surface becomes further inclined to the flow, the velocity of the reflected shock increases until a stage is eventually reached at which the point of intersection is forced to leave the wall. A third shock, known as the *Mach stem*, perpendicular to the wall, forms at the point of intersection which is now termed the *triple point*. Examination of typical particle paths shows that the gas immediately adjacent to the wall passes through the single shock while that further away is processed by two shocks. This creates a mismatch in the velocities of the shock-heated gas streams and leads to a *slip line*. This behaviour, illustrated in Fig. 4.10(c) is called *Mach reflection*.

We are now in a position to describe a simple *cellular detonation*, although we shall only attempt a two-dimensional representation as this is easier to follow. The detonation comprises alternating convex shocks and Mach stems connected by triple points as described above. Leading back from each triple point is a reflected shock termed here a *transverse wave*. As the detonation front propagates the transverse waves move into the regions processed by the curved incident shock. When the transverse waves collide, the pattern is effectively inverted so that the rapid reaction initiated by the collison leads to a new Mach stem and a new pair of transverse waves moving outwards into the now weakened incident shock.

This complex wave behaviour can be studied relatively easily because the velocity shear associated with the slip at the triple point causes it to 'write' on a soot-coated surface leading to the type of pattern illustrated in Fig. 4.11. The transverse wave spacing and the cell length are smaller the stronger the detonation: while the relationship is not fully understood, it seems that the cell structure is coupled to the reaction zone length and the geometry of the tube.

In view of this complex three-dimensional structure one would

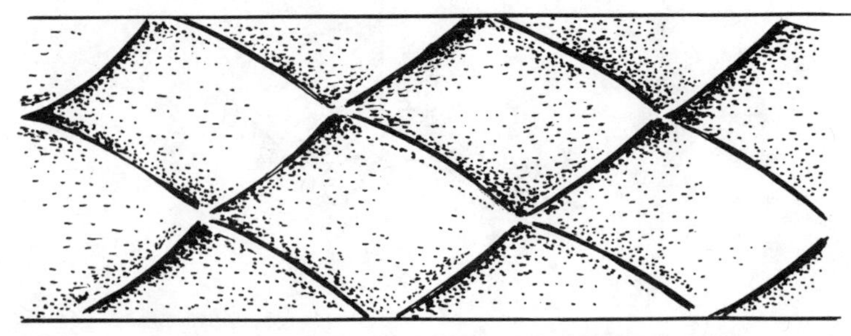

Fig. 4.11 Smoke track records showing cellular detonation structure (after Strehlow [89]).

expect the one-dimensional C–J theory to be totally invalid [90]. Obviously, the three-dimensional structure must average out to the C–J behaviour in a way that is not yet completely understood. Certainly C–J calculations are universally employed and are often more reliable than experimental measurements.

4.5 INITIATION OF DETONATION

For the majority of exothermic reaction mixtures, two quite stable combustion waves are possible, a propagating flame and a detonation. Because of the potential of the detonation for causing severe damage, it is of general interest to attempt to understand how detonation may be initiated. Three distinct modes of initiation need to be considered – by shock wave, by electric spark, and by flame. In all cases, the degree of confinement or the geometry of the container play a dominant role.

Perhaps the most straightforward situation is typified by a shock wave reflecting from the closed end of a tube. Behind the shock wave there is an induction zone within which there is little change in the gross properties of the system, although branching-chain reactions will be occurring. This zone is followed by the liberation of heat to the gas as the chemical reactions approach completion. Thus a *reaction wave* follows the shock wave at a constant distance from it. However, at the end wall the release of energy causes an adiabatic explosion in the stationary gas which generates a *reaction shock wave* which travels in the induction zone ahead of the reaction wave. The effect of the reaction shock is to reduce the length of the induction zone so that the two waves travel together until they overtake the original reflected shock at which the steady detonation forms. It should be appreciated that the reaction wave is not really a flame, although that name is sometimes used, because propagation does not involve transport processes.

Although the model is reasonably simple in the case of the reflected shock wave, the general principle will be the same for other shock waves. Thus a reaction wave of the type described above will always follow a shock wave in a combustible mixture. Any perturbation associated with this reaction wave which leads to a weak shock or compression wave can then cause the reaction wave to accelerate and eventually overtake the front.

Initiation by electric spark is interesting because it can lead either to a normal pre-mixed flame or to a detonation depending on conditions. The reason is that the electric spark generates a spherical shock wave followed by a *kernel* of spark-heated gas. If the temperature and pressure behind the expanding shock wave remain sufficiently high to cause rapid reaction, i.e. to generate a reaction wave, then detonation can follow from the mechanism described above. If this condition is not met, then it is still possible for the hot kernel to ignite a flame by heat transfer to the surrounding gas.

Initiation by flame is the most complex process of all. Once again, the behaviour depends critically on the nature of the container. When the flame is ignited at the closed end of a long tube, the transition to detonation often occurs only at the end of a long *pre-detonation run.* From Fig. 4.6, it is apparent that the burnt gas behind a flame has a lower density, and hence occupies a larger volume, than that of the gas ahead. Thus the flame front leaving the closed end of the tube will travel at several times the laminar burning velocity and will generate pressure waves in the unburnt gas ahead. These cause pre-heating of the gas before it enters the flame and hence when the gas actually enters the flame even more rapid reaction ensues. The planar flame front soon becomes unstable and turbulent motion further increases the rate of burning. Pockets of unburnt gas become trapped in the burning zone and are compressed and heated there until self-ignition occurs and a detonation eventually forms.

4.6 SUGGESTIONS FOR FURTHER READING

Bradley, J.N. (1962) *Shock Waves in Chemistry and Physics*, Methuen, London.

Cook, M.A. (1958) *The Science of High Explosives*, Reinhold, New York.

Lewis, B. and Von Elbe, G. (1961) *Combustion, Flames and Explosions of Gases*, 2nd edn, Academic Press, New York.

Shchelkin, K.I. and Troshin, Y.A.K. (1965) *Gasdynamics of Combustion*, Mono Book Corp., Baltimore.

Soloukhin, R.I. (1965) *Shock Waves and Detonations in Gases*, Mono Book Corp., Baltimore.

Strehlow, R.A. (1979) *Fundamentals of Combustion*, Kreiger, Huntingdon, New York.

Fickett, W. and Davies, W.C. (1979) *Detonation*, University of California Press, Berkeley.

4.7 PROBLEM

Use the approximate formulae to estimate the detonation temperature, velocity and pressure for stoichiometric hydrogen–air and methane–air mixtures initially at standard temperature and pressure.

5

The chemistry of combustion

5.1 DESCRIPTION OF IMPORTANT REACTIONS

In the next three chapters, we shall be primarily concerned with the chemical processes involved in combustion. Combustion phenomena occur because the overall chemical reaction is exothermic: however, the nature and character of the events depend to some extent on the details of the reactions involved, on the mechanism of the reaction, and on the rates of the elementary steps. For example, in a branching-chain reaction there is generally a delicate balance between radical multiplication by chain branching and radical destruction by chain-termination processes. In these circumstances, the whole behaviour of the system may be changed from slow reaction to explosion, or vice versa, by quite small alterations in the rates of individual reaction steps. Although branching-chain reactions are not essential for the propagation of flames, the reactions involved are necessarily rapid and usually involve chain processes of some sort. Until fairly recently, theories of flame propagation intended to give quantitative comparison with experiment commonly employed global, or overall, kinetics, that is, rate expressions based on a relationship of the form

$$\frac{-\mathrm{d}[\text{fuel}]}{\mathrm{d}t} = A[\text{fuel}]^{B}[\text{oxygen}]^{C}\exp(-E/RT) \qquad (5.1)$$

where the parameters A, E, B and C are established empirically. Because such relationships make no allowance for the detailed chemistry, they are able to explain the observed flame characteristics only over a very limited range of conditions. A much more satisfactory approach has become possible with the advent of powerful computers and now, as explained in Chapter 3, the modern analysis of flame properties takes full account of the flame reactions.

Table 5.1 Arrhenius parameters for some bimolecular reactions [91, 92]

Reaction	$A(\mathrm{dm^3\,mol^{-1}\,s^{-1}})$	$E(\mathrm{kJ\,mol^{-1}})$
$H + O_2 \rightarrow OH + O$	2.24×10^{11}	70.3
$O + H_2 \rightarrow OH + H$	1.74×10^{10}	39.5
$OH + H_2 \rightarrow H_2O + H$	2.2×10^{10}	21.5
$O + OH \rightarrow O_2 + H$	1.3×10^{10}	0
$O + H_2O \rightarrow OH + OH$	5.75×10^{10}	75.3
$H + H_2O \rightarrow H_2 + OH$	8.41×10^{10}	84.1
$H + CH_4 \rightarrow H_2 + CH_3$	1.3×10^{11}	49.8
$CH_3 + CH_3OH \rightarrow CH_4 + CH_3O$	2.0×10^{11}	41.0
$CH_3 + CH_3CHO \rightarrow CH_4 + CH_3CO$	1.4×10^{12}	35.3

To be of importance in combustion, a reaction must be reasonably rapid. In general, appreciable amounts of reaction must occur in times within the range 10^{-6} to 10^{-3} s (1 μs to 1ms). The effect of temperature on a rate constant is normally given by an Arrhenius equation

$$k = A\exp(-E/RT) \tag{5.2}$$

in which the pre-exponential factor A is independent of, or varies only slowly with, temperature while the activation energy, E, is a constant. Since combustion reactions take place at high temperatures, reactions with a high activation energy and high pre-exponential factor tend to be more important than those with a lower activation energy but lower pre-exponential term. This is in complete contrast with the behaviour at room temperature. Another feature worth mentioning is that many of the simpler bimolecular reactions have very similar pre-exponential factors and differ only in their activation energies (Table 5.1). Since the exponential term approaches unity at high temperatures, many of these reactions turn out to have similar rate constants at flame temperatures and hence become of comparable importance.

Because most oxidations involve two components, a fuel and an oxidant, unimolecular reactions are normally important only for a limited number of self-decomposition reactions. As a rough generalization it may be said that unimolecular reactions involving molecules have high activation energies ($> 250\,\mathrm{kJ\,mol^{-1}}$); labile molecules such as peroxides are an exception and free radicals also decompose by lower activation energy paths (Table 5.2).

Among bimolecular reactions, the most important ones are metathetical reactions involving an atom or simple radical. Such reactions do not usually involve complex rearrangements and therefore have

Table 5.2 Arrhenius parameters for some unimolecular reactions [93, 94].

Reaction	$A(s^{-1})$	$E(kJ\,mol^{-1})$
$C_2H_6 \rightarrow 2CH_3$	5.0×10^{16}	370
$n\text{-}C_4H_{10} \rightarrow 2C_2H_5$	1.9×10^{17}	342
$i\text{-}C_4H_{10} \rightarrow CH_3 + i\text{-}C_3H_7$	6.3×10^{17}	345
$(CH_3)_4C \rightarrow CH_3 + t\text{-}C_4H_9$	6.3×10^{16}	343
$C_3H_6 \rightarrow CH_3 + C_2H_3$	1.3×10^{16}	359
t-BuOOH → t-BuO + OH	4×10^{15}	176
$n\text{-}C_3H_7 \rightarrow C_2H_4 + CH_3$	1.6×10^{14}	136
$n\text{-}C_4H_9 \rightarrow C_2H_4 + C_2H_5$	2.5×10^{13}	120
$s\text{-}C_4H_9 \rightarrow C_3H_6 + CH_3$	7.3×10^{14}	137
$i\text{-}C_4H_9 \rightarrow C_3H_6 + CH_3$	2.8×10^{12}	130

high pre-exponential factors. The most common of reactions in this category are *abstractions*, for example,

$$H + CH_4 \rightarrow H_2 + CH_3 \tag{5.3}$$

$$OH + CH_4 \rightarrow H_2O + CH_3 \tag{5.4}$$

Furthermore, their activation energies are not as high as those in unimolecular reactions which involve bond dissociation and they are therefore expected to predominate, especially at lower temperatures. The nature of the species involved clearly depends on the particular fuel–oxidant system but in virtually all systems in which the elements H and O occur, the OH radical is an important and frequently dominant species causing oxidation, although reactions involving HO_2, H and O are also significant in many cases. The rate of attack by such a species on a particular alkane can be obtained from an additivity rule by assuming that the contribution per C–H bond to the total rate constant is the same for all primary, all secondary, and all tertiary bonds in each hydrocarbon. The overall rate constant is then given by Equation 5.5,

$$k = n_p A_p \exp(-E_p/RT) + n_s A_s \exp(-E_s/RT) + n_t A_t \exp(-E_t/RT) \tag{5.5}$$

where n is the number of bonds of a specific type, A is the Arrhenius pre-exponential factor per C–H bond, and E is the corresponding activation energy.

The subscripts p, s, t, refer to attack at primary, secondary and tertiary C–H bonds, respectively. Table 5.3 gives values of the Arrhenius parameters for several radical–alkane reactions of importance in hydrocarbon oxidation.

Table 5.3 Values of A(per C–H bond) and E for radical attack at primary, secondary and tertiary bonds [93, 95, 96, 97].

Reaction	Primary		Secondary		Tertiary	
	A(dm^3 mol^{-1} s^{-1})	E(kJ mol^{-1})	A(dm^3 mol^{-1} s^{-1})	E(kJ mol^{-1})	A(dm^3 mol^{-1} s^{-1})	E(kJ mol^{-1})
H + RH	2.2×10^{10}	39.2	4.9×10^{10}	33.3	5.1×10^{10}	25.2
O + RH	5.0×10^{9}	24.2	1.3×10^{10}	18.8	1.6×10^{10}	13.8
OH + RH	3.9×10^{7}	28.3	3.1×10^{6}	22.7	4.9×10^{7}	20.1
HO_2 + RH	4.9×10^{7}	62.5	4.9×10^{7}	52.6	4.9×10^{7}	41.5
RO_2 + RH	4.9×10^{7}	62.5	4.9×10^{7}	52.6	4.9×10^{7}	41.5
CH_3 + RH	4.9×10^{8}	49.0	3.3×10^{8}	42.3	2.4×10^{8}	33.6
CH_3O + RH	5.3×10^{7}	29.5	3.6×10^{7}	18.7	1.9×10^{7}	11.6

Another class of bimolecular reactions whose importance in combustion systems has recently been recognized involves two radical species. A *radical–radical* reaction may cause chain termination, as when two alkyl radicals combine to form an alkane, for example,

$$2C_2H_5 \rightarrow C_4H_{10} \tag{5.6}$$

On the other hand, radical–radical reactions may result in chain propagation, for example,

$$2CH_3O_2 \rightarrow 2CH_3O + O_2 \tag{5.7}$$

the alkoxy radicals undergoing further reaction to products such as aldehydes and alcohols.

The rate constants of radical–radical reactions are normally large, approaching the collision frequency and their activation energies are very small or zero: consequently their rates vary very little with temperature over the range of interest. Values of some rate constants of important radical–radical reactions are given in Table 5.4.

In many systems, radical concentrations are low, and therefore radical–radical reactions whose rates depend on the product of two radical concentrations are unimportant except for termination processes. However, in combustion where rates of reaction are high, radical concentrations may rise appreciably and the rate of radical–radical reactions may no longer be negligible compared with radical–molecule reactions which possess a significant activation energy.

Termolecular reactions are rare in most chemical systems, apart from an important class of atom and radical recombination reactions in which a 'third body' or 'chaperon' is required to remove the energy released

$$X + X + M \rightarrow X_2 + M \tag{5.8}$$

The rate constants for most recombination reactions fall in the range

Table 5.4 Rate constants for some important radical–radical reactions [94].

Reaction	$k(\mathrm{dm^3\,mol^{-1}\,s^{-1}})$
$CH_3 + CH_3 \rightarrow C_2H_6$	2.6×10^{10}
$C_2H_5 + C_2H_5 \rightarrow C_4H_{10}$	1×10^{10}
$CH_3 + O \rightarrow HCHO + H$	1×10^{11}
$2CH_3O_2 \rightarrow 2CH_3O + O_2$	8.7×10^7
$2CH_3O_2 \rightarrow CH_3OH + HCHO + O_2$	1.5×10^8

10^8 to 10^{10} $dm^6\,mol^{-2}\,s^{-1}$ and vary by a factor of up to 20, depending on the third body.

Although processes such as Reaction 5.8 have a negligible or even negative activation energy, the number of three-body collisions is so low that these reactions might be expected to be unimportant. However, since they provide virtually the only route available for the homogeneous removal of reactive species in gaseous combustion and, at the same time, are responsible for the liberation of considerable amounts of energy (in many instances 75% of the available enthalpy appears as a result of such processes) they are very important. For example, such reactions tend to predominate in the post-flame region which follows the main reaction zone of a flame.

The reactive species involved in combustion reactions are normally atoms or radicals, although there are instances in which labile molecules or electronically excited species play a similar role. Particularly in flames, only simple radicals are important, more complex species usually being too labile and decomposing rapidly to give simpler entities.

The chemistry of combustion will now be illustrated in greater detail by considering two important and extensively studied reactions: the oxidation of hydrogen and the oxidation of carbon monoxide. These reactions are more fully understood than any others and they provide examples of many of the phenomena already described.

5.2 THE HYDROGEN–OXYGEN REACTION

Much more is known about the hydrogen–oxygen system than any other comparable combustion reaction and it is normally used as a model for discussing the behaviour of branching-chain reactions (Chapter 2). Despite prolonged argument over details the major reactions are now fully understood [98, 99].

5.2.1 The basic mechanism

Pre-mixed flames of hydrogen in oxygen or air are quite typical except that they show little or no visible radiation, the amount which is observed normally being due to trace impurities. Considerable amounts of OH radiation can however be detected in the ultraviolet region of the spectrum. For a stoichiometric hydrogen–oxygen flame the temperature is about 3100 K and the burnt gas composition shows 57% conversion to water with about one-quarter of the gas remaining as radicals H, O and OH [9].

In static systems, no reaction occurs below 400° C unless the mixture is ignited by an external source such as a spark, whilst above 600° C explosion occurs spontaneously at all pressures. Between these two temperatures, three separate explosion limits can be identified (see Fig. 2.8), the first limit occurring at pressures of up to a few torr and the second at pressures of up to 200 torr.

The basic chain sequence is composed of three reactions, two of which are branching:

$$H + O_2 \rightarrow OH + O \qquad \text{Branching} \qquad (5.9)$$

$$O + H_2 \rightarrow OH + H \qquad \text{Branching} \qquad (5.10)$$

$$OH + H_2 \rightarrow H_2O + H \qquad \text{Propagating} \qquad (5.11)$$

A complete cycle commences with one hydrogen atom and leads to the production of three hydrogen atoms according to the stoichiometric relation

$$H + 3H_2 + O_2 \rightarrow 3H + 2H_2O \qquad (5.12)$$

Reaction 5.9 has an activation energy of $70 \, kJ \, mol^{-1}$, just slightly greater than its endothermicity; the activation energies of Reactions 5.10 and 5.11 are considerably smaller and so the rate of Reaction 5.9 is appreciably lower than that of Reactions 5.10 and 5.11. At temperatures less than about 1200 K much of the mathematical analysis has been based on the assumption that Reaction 5.9 is rate determining so that the OH and O concentrations remain in steady state with the H atom concentration throughout. This type of analysis is common in branching-chain reactions although it must be used with caution [100]. Certainly at high temperatures the rate of Reaction 5.10 can become of almost equal importance to the rate of Reaction 5.9.

These three reactions are responsible for the exponential increase in rate when explosion occurs. However, they do not correspond to the correct overall stoichiometry

$$2H_2 + O_2 \rightarrow 2H_2O \qquad (5.13)$$

and other reactions must become important as reaction proceeds. In flames, for example, the branching chain reaction ceases when the reverse processes to Reactions 5.9, 5.10 and 5.11 become significant, that is,

$$OH + O \rightarrow H + O_2 \qquad (5.14) \equiv (-5.9)$$

$$OH + H \rightarrow O + H_2 \qquad (5.15) \equiv (-5.10)$$

$$H + H_2O \rightarrow OH + H_2 \qquad (5.16) \equiv (-5.11)$$

A quasi-equilibrium is then established in which the reaction has proceeded perhaps three-quarters of the way to completion but a large amount of the available enthalpy is still contained in the high atom and radical concentrations. This energy is then released by third-order recombination reactions which take place on a longer time scale than the processes above. In the post-flame gases of fuel-rich mixtures the dominant reactions are

$$H + OH + M \rightarrow H_2O + M \tag{5.17}$$

$$H + H + M \rightarrow H_2 + M \tag{5.18}$$

while in the fuel-lean case, processes involving the HO_2 radical, for example,

$$OH + HO_2 \rightarrow H_2O + O_2 \tag{5.19}$$

$$H + HO_2 \rightarrow H_2 + O_2 \tag{5.20}$$

become important.

The different nature of the two parts of the flame is revealed in other ways: in the luminous reaction zone, the high radical concentrations manifest themselves by non-equilibrium emission and ion concentrations above equilibrium levels, whereas in the post-flame gases heat is still evolved but the radiation corresponds closely to thermal equilibrium.

5.2.2 Explosion limits

Reverting now to a consideration of the reaction in a static system, at pressures below the first limit the chain reaction is so slow that it is not normally detectable. The first limit pressure decreases as the vessel diameter is increased; it is dependent on the nature of the vessel surface and the limit pressure may be reduced so that explosion is facilitated by the addition of inert gas. These observations point to the destruction of active centres at the surface. The limit is markedly dependent on the oxygen concentration in the reactants, but almost independent of the hydrogen concentration. This suggests that the competition is between the reaction of an active centre with oxygen giving propagation or branching, and the destruction of that centre at the wall. The active centre concerned must be the H atom since this is the only species which reacts with oxygen in the main cycle comprising Reactions 5.9, 5.10 and 5.11. This conclusion is not entirely unexpected since the diffusion coefficient of H atoms is much higher than that of any other active centre in the system and it is therefore the chain carrier most likely to be removed at the wall. More detailed analysis shows that for a precise interpretation of the first limit, allowance must also be made for smaller contributions from the destruction of O atoms at the wall and from homogeneous termination. When all these factors are included, the observed limit can be predicted to within $\pm 1\%$ [36]. At the first limit, the multiplication of chain carriers by branching (Reaction 5.9) is just held in check by removal of H atoms at the wall; thus, if the diffusion coefficient of H atoms through the reaction mixture can be calculated, the rate constant of the branching reaction may be obtained [101].

At pressures above the first limit, the reaction remains explosive until the second limit is reached. The second limit pressure is only slightly affected by changes in vessel size or surface while the addition of inert gas, in contrast to the behaviour at the first limit, raises the limit pressure, that is, makes the mixture less explosive. These observations point to the reaction which controls branching operating in the gas phase. This reaction which competes with branching is

$$H + O_2 + M \rightarrow HO_2 + M \tag{5.21}$$

The HO_2 radical is relatively unreactive and is therefore much more likely than the other active centres H, O and OH, to diffuse to the walls and be deactivated there. In this case, Reaction 5.21 simply acts as a chain-termination process.

At the second limit, provided all the HO_2 radicals are destroyed in this way, the rate of production of radicals is just balanced by their destruction in Reaction 5.21 and thus

$$2k_9[H][O_2] = k_{21}[H][O_2][M] \tag{5.22}$$

where k_9 and k_{21} are the rate constants of Reactions 5.9 and 5.21, respectively, though it must be noted that the efficiency of molecules as 'chaperons' in Reaction 5.21 varies considerably. From Equation 5.22 it follows that

$$[M] = 2k_9/k_{21} \tag{5.23}$$

and so the temperature dependence of the second limit depends on $E_9 - E_{21}$.

The [M] term includes a contribution from each of the molecular species present. Since at a fixed temperature, the concentration of a gas is proportional to its partial pressure

$$p_l = p_{H_2} + \beta_{O_2} p_{O_2} + \beta_X p_X \tag{5.24}$$

where p_l is the total pressure at the limit, X is any inert gas present and the β values refer to the effectiveness of the species relative to hydrogen in stabilizing HO_2 by collision. As mentioned in Chapter 2, the value of β_X, at least for simple molecules such as O_2 and N_2, is accurately predicted by the collision frequency for $HO_2 + X$ divided by that for $HO_2 + H_2$.

If the pressure is increased above the second limit, the HO_2 radical is prevented from diffusing to the walls and can then propagate the chain by

$$HO_2 + H_2 \rightarrow H_2O_2 + H \quad (5.25)$$

$$HO_2 + H_2 \rightarrow H_2O + OH \quad (5.26)$$

of which the former seems more probable. Certainly hydrogen peroxide has been detected in the products of reaction [102] but under conditions favouring high concentrations of HO_2 the reaction

$$HO_2 + HO_2 \rightarrow H_2O_2 + O_2 \quad (5.27)$$

also takes place. Since hydrogen peroxide dissociates relatively easily

$$H_2O_2 (+M) \rightarrow 2OH\ (+M) \quad (5.28)$$

the HO_2 radical can also lead to additional degenerate branching under these circumstances.

The third explosion limit which is observed at pressures of several hundred torr was originally believed to be due to Reaction 5.25 which effectively regenerates an active centre (H) from the relatively inert HO_2. Consequently, the explosive reaction which had been quenched by Reaction 5.21 was rekindled. This explanation is considerably over-simplified: first, the reactions of H_2O_2 are neglected and secondly, just below the limit, there is considerable self-heating of the reaction mixture [43, 44]. While the first and second limits can be regarded as virtually isothermal, a full understanding of the third limit can only be reached with the aid of the unified chain-thermal theory [103, 104].

No mention has been made so far of the reactions which lead to chain initiation. The main reason for this is that such reactions have little effect on the overall behaviour and kinetics, which means they are difficult to elucidate. Under some circumstances, initiation may well be heterogeneous, but at high temperatures the bimolecular gas-phase exchange reaction

$$H_2 + O_2 \rightarrow 2OH \quad (5.29)$$

provides the initial supply of radicals [105].

5.2.3 Effect of surfaces

The nature of the vessel surface plays an important role in all aspects of the hydrogen–oxygen reaction. The precise mechanisms involved are not properly understood but at low pressures, in the vicinity of the first limit, the rate-determining step probably involves the adsorption of chain carriers (principally H atoms) followed by recombination to give stable molecules which are then desorbed. Treatment of the walls by surface coating or by 'ageing' (the performance of successive experiments in the same vessel) has a drastic effect on the kinetics of the reaction presumably because it modifies the efficiency of the chain-breaking step. Thin coats of various salts on the surface can cause up to a one-hundred-fold variation in the pressure at the first explosion limit. Surfaces coated with KOH, CsCl, Al_2O_3 and K_2HPO_4 are highly efficient at removing chain carriers, and in vessels with these surfaces first explosion limit pressures are high. On the other hand, B_2O_3-coated silica and acid-washed pyrex surfaces are very inefficient while those coated with KCl, NaCl, $BaCl_2$ and $Na_2B_4O_7$ are of intermediate efficiency.

At higher pressures, the kinetics of reaction depend on the fates of the HO_2 radical and of H_2O_2, formed by gas-phase reactions of HO_2, at the vessel walls. On some surfaces HO_2 reacts to give hydrogen peroxide which is released into the gas phase, whilst on others it reacts to give oxygen and water. The effect on the kinetics will then differ because H_2O_2 can participate directly in the chain reaction. A corresponding situation occurs for H_2O_2 itself which is unaffected on some surfaces but reacts to give oxygen and water on others.

A classification of surfaces has been proposed [106] as follows:

Class I. Acid surfaces catalyse the reaction by proton donation and release hydrogen peroxide and oxygen.

$$2HO_2 \xrightarrow{H^+} H_2O_2 + O_2 \tag{5.30}$$

The reaction is slow relative to diffusion and is therefore rate controlling. Hydrogen peroxide itself is unaffected.

Class II. Metal oxides and salts destroy both HO_2 and H_2O_2 but with much greater efficiency towards HO_2 than H_2O_2. The effect on the kinetics of reaction is that the loss of HO_2 is purely diffusion controlled, whilst the surface reaction controls the rate of destruction of H_2O_2.

Class III. Strong electron donors, for example, metals such as silver and gold, destroy both species with high efficiency. The stoichiometry is represented by

$$4HO_2 \xrightarrow{e^-} 2H_2O + 3O_2 \tag{5.31}$$

$$2H_2O_2 \xrightarrow{e^-} 2H_2O + O_2 \tag{5.32}$$

Combustion is clearly most sensitive to diffusion control with such surfaces.

Very similar behaviour towards surfaces is observed in other systems, for example, the oxidation of methane, and this can throw some light on the gas-phase processes occurring in those systems.

5.2.4 The slow reaction

At pressures above the second explosion limit the reaction proceeds with measurable but increasing rapidity until the third explosion limit. The rate of reaction is affected considerably by the nature of the surface and only in aged boric acid-coated vessels is it sufficiently reproducible for a comprehensive mechanism to have been established. This mechanism gives a quantitative account not only of the slow reaction itself but also of the induction period and second limit behaviour in aged boric acid-coated vessels [98].

The important reactions under these conditions are given in Table 5.5. It is also necessary to include an initiating step, though its

Table 5.5 Elementary steps in the slow reaction between hydrogen and oxygen in aged boric acid-coated vessels at temperatures around 500° C.

$OH + H_2 \rightarrow H_2O + H$
$H + O_2 \rightarrow OH + O$
$O + H_2 \rightarrow OH + H$
$H + O_2 + M \rightarrow HO_2 + M$
$H_2O_2 + M' \rightarrow 2OH + M'$
$H + HO_2 \rightarrow 2OH$
$2HO_2 \rightarrow H_2O_2 + O_2$
$HO_2 + H_2 \rightarrow H_2O_2 + H$
$H + H_2O_2 \rightarrow H_2O + OH$
$H + H_2O_2 \rightarrow H_2 + HO_2$
$OH + H_2O_2 \rightarrow H_2O + HO_2$

nature need not be specified. As the surface ages, it becomes increasingly inert towards HO_2 and H_2O_2, and eventually reaches a state where the surface reactions of these species are unimportant. The decomposition of the reaction product, hydrogen peroxide, augments the supply of radicals by the primary initiation step, and production of radicals by this secondary initiation (or degenerate branching) process becomes of increasing importance as the reaction develops.

Now that the slow reaction is so well understood under these conditions, it is possible to use a reacting mixture of hydrogen and oxygen as a source of controlled concentrations of H, OH and HO_2. In this way, much information has been gained on the rate constants for attack by these species on additives such as hydrocarbons thereby adding greatly to our understanding of hydrocarbon combustion (Chapter 6).

5.3 THE OXIDATION OF CARBON MONOXIDE

After the oxidation of hydrogen, this is probably the best understood example of a branching-chain reaction [99]. It is of great importance because the initial stage in the oxidation of organic compounds leads to the formation of carbon monoxide which is then further oxidized to carbon dioxide if sufficient oxygen is present.

A most interesting feature of carbon monoxide flames is their sensitivity to trace amounts of water or hydrogen. It is quite difficult to produce a CO/O_2 flame if the gases are completely dry but a small trace of water has a dramatic effect on the flame speed (0.25% raises the velocity from 1 to 7.8 m s^{-1}). The burning velocity is proportional to the square root of the CO concentration and from the discussion on flame propagation this indicates that the reaction is proportional to [CO] and independent of $[O_2]$. In a stoichiometric mixture, the flame temperature is calculated to be 2973 K and the equilibrium burnt gas composition still contains 35% of unchanged carbon monoxide.

In a static system, two limiting peninsulae are observed, one corresponding to light emission or 'glow' and the other to a drastic pressure increase or 'explosion' (Fig. 5.1). The two curves coincide at the second limit but are quite well separated at the first limit. The behaviour of this system, as with all branching-chain systems, depends on the nature of the vessel surface and on its geometry. The sensitivity of the reaction to trace amounts of moisture, which is even more critical, is not so typical of branching-chain reactions.

The glow limit corresponds to a transition from a slow to a rapid

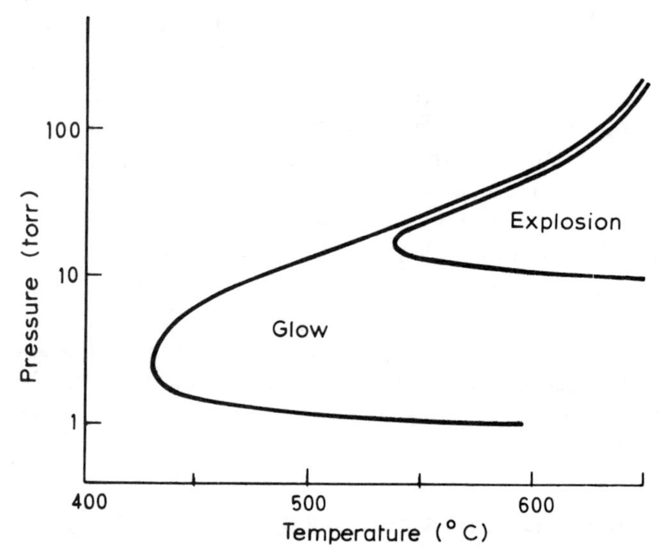

Fig. 5.1 Explosion boundaries for the carbon monoxide–oxygen system (after Hoare and Walsh [107]).

reaction rate in much the same manner as the explosion limit and the first phenomenon which requires explanation is the occurrence of two such transitions. The general characteristics of first explosion limits were discussed in some detail in Chapter 2 and it is found that the same characteristics, notably the effects of geometry and of added inert gases, are observed for the glow limit rather than the explosion limit. It seems therefore that the first transition corresponds to the net branching factor becoming positive and is a true chain-isothermal boundary but that the reaction is self-inhibiting so that explosion cannot occur. The inhibition is presumably caused by a product formed during the reaction. The self-inhibiting effect can be demonstrated by careful rate measurements and is particularly well illustrated if the system is brought very rapidly to the glow limit, so that a negligible extent of reaction occurs in the slow reaction region, as a result of which a true explosion ensues.

The explosion limit then probably corresponds to a chain-thermal explosion rather than to a 'normal' branching-chain explosion. Since an increase in pressure causes an increase in reaction rate and hence in heat output, while at the same time reducing heat loss to the walls, a stage is reached at which the inhibiting effect of the products is inadequate to prevent the acceleration due to self-heating.

Confirmation of this notion is provided by the fact that added argon lowers the limit, whilst nitrogen with its higher thermal conductivity raises it.

For a typical chain reaction, two active centres are required to provide a 'shuttle' (Chapter 2). It is agreed that one of the active centres in the CO–O_2 system must be the oxygen atom but the identity of the other is not so certain. Both ozone and an electronically excited state of carbon dioxide have been proposed, the balance of the evidence now supporting the latter. Certainly, the glow itself arises from an electronically excited state of carbon dioxide. The mechanism is based on an initiation step

$$CO + O_2 \rightarrow CO_2 + O \tag{5.33}$$

followed by propagation

$$O + CO \rightarrow CO_2^* \tag{5.34}$$

and branching

$$CO_2^* + O_2 \rightarrow CO_2 + 2O \tag{5.35}$$

The excited CO_2 is probably in the $^3B_2$ state, in which case Reaction 5.35 is 33 kJ mol^{-1} exothermic. Propagation may also occur when excited CO_2 reacts with CO.

$$CO_2^* + CO \rightarrow CO_2 + CO^* \tag{5.36}$$

followed by

$$CO^* + O_2 \rightarrow CO_2 + O \tag{5.37}$$

In contrast to the hydrogen–oxygen reaction, this mechanism provides an example of an energy-branching chain.

At low pressures, chains are terminated at the wall by removal of O and also perhaps CO_2^*. The upper limit, as before, corresponds to a chain-termination process which must be of higher kinetic order than the propagation reaction; this termination process is probably

$$CO + O + M \rightarrow CO_2 + M \tag{5.38}$$

The inhibition of explosion at the glow limit could be attributed either to a major product or to some species formed in low amounts. If a major product were responsible, this would necessarily be carbon dioxide and the inhibiting reaction could then be

$$O + CO_2 \rightarrow O_2 + CO \tag{5.39}$$

Evidence has also been provided for the formation of small amounts of carbon and carbon suboxides in the system possibly by such reactions as

$$CO^* + CO \begin{cases} \nearrow CO_2 + C \\ \searrow C_2O + O \end{cases} \qquad (5.40)$$

$$C_2O + CO \rightarrow C_3O_2 \qquad (5.41)$$

These suboxides could inhibit the reaction by removing oxygen atoms, hence reducing the rate of the branching reaction, for example,

$$C_2O + O \rightarrow 2CO \qquad (5.42)$$

$$C_3O_2 + O \rightarrow 3CO \qquad (5.43)$$

$$C_3O_2 + O \rightarrow C_2O + CO_2 \qquad (5.44)$$

The different behaviour observed when the system is brought rapidly or slowly to the critical condition is explained better by this mechanism than by CO_2 inhibition.

The effect of trace quantities of moisture is not fully understood although many features of the reaction, and in particular the position of the second limit, are considerably affected. Hydrogen has a similar effect and the addition of as little as 0.1% widens the ignition limits considerably and extends the explosion peninsula to lower temperatures. When about 1% is added, the explosion peninsula is virtually the same as that of the hydrogen–oxygen system and when 10% of the carbon monoxide is replaced by hydrogen, the behaviour at the second limit is almost identical to that of an equivalent H_2–O_2–N_2 mixture. One explanation is that additional branching occurs via the reaction

$$CO + O + H_2O \rightarrow CO_2 + H + OH \qquad (5.45)$$

This third-order process competes with

$$CO + O + M \rightarrow CO_2 + M \qquad (5.38)$$

and leads to further branching rather than inhibition. Subsequent processes will include

$$OH + CO \rightarrow H + CO_2 \qquad (5.46)$$

and reactions typical of the hydrogen–oxygen system, for example,

$$H + O_2 \rightarrow OH + O \qquad (5.9)$$

The carbon monoxide–oxygen reaction also exhibits oscillatory behaviour under some circumstances: this will be discussed in Chapter 7 when some specialized aspects of combustion reactions are dealt with.

5.4 OTHER OXIDATION REACTIONS

The mechanisms of oxidation of other inorganic compounds are far less well understood, mainly because of their greater complexity. However, evidence [100] is now accumulating for a common pattern of behaviour in which the primary chain is rapid and non-branching, usually involving the hydroxyl radical. A product of appreciable stability, which can lead to branching at high temperatures, is produced in this primary chain. Thus, in the hydrogen sulphide–oxygen system, the following sequence is believed to occur.

$$\left.\begin{aligned} OH + H_2S &\rightarrow H_2O + HS \\ HS + O_2 &\rightarrow SO + OH \end{aligned}\right\} \text{Primary chain} \qquad (5.47),\ (5.48)$$

$$\left.\begin{aligned} SO + O_2 &\rightarrow SO_2 + O \\ O + H_2S &\rightarrow OH + HS \end{aligned}\right\} \text{Branching cycle} \qquad (5.49),\ (5.50)$$

The role of the species SO, which is a molecular intermediate with a longer lifetime than the radicals present, is probably performed by HNO in the ammonia–oxygen system. Here, the fuel has an inhibiting effect because it competes with oxygen for the H atoms present

$$H + NH_3 \rightarrow NH_2 + H_2 \qquad (5.51)$$

The resulting radical reacts with oxygen to propagate the chain, at the same time producing a molecular intermediate with a longer lifetime which is eventually responsible for branching

$$NH_2 + O_2 \rightarrow HNO + OH \qquad (5.52)$$

$$HNO \rightarrow H + NO \qquad (5.53)$$

The reaction

$$OH + NH_3 \rightarrow H_2O + NH_2 \qquad (5.54)$$

completes the cycle. This behaviour is closely parallel to the degenerate branching observed with hydrogen peroxide in the hydrogen–oxygen slow reaction. It also strongly resembles the behaviour of organic molecules described in the next chapter, although the characteristics of low-temperature hydrocarbon oxidation have not been observed in inorganic systems.

5.5 SUGGESTIONS FOR FURTHER READING

Baldwin, R.R. and Walker, R.W. (1972) Branching-chain reactions: the hydrogen–oxygen reaction, in *Essays in Chemistry* (ed. J.N. Bradley, R.D. Gillard and R.F. Hudson), Vol. 3, Academic Press, London, p. 1.

Dainton, F.S. (1966) *Chain Reactions, An Introduction*, 2nd edn, Methuen, London.

Dixon-Lewis, G. and Williams, D.J. (1977) The oxidation of hydrogen and carbon monoxide, in *Comprehensive Chemical Kinetics* (ed. C.H. Bamford and C.F.H. Tipper) Vol. 17, Elsevier, Amsterdam, p. 1.

Hinshelwood, C.N. (1940) *The Kinetics of Chemical Change*, Oxford University Press, Oxford.

Kondratiev, V.N. (1964) *Chemical Kinetics of Gas Reactions*, Pergamon, London.

Lewis, B. and Von Elbe, G. (1961) *Combustion, Flames and Explosions of Gases*, 2nd edn, Academic Press, New York.

Minkoff, G.J. and Tipper, C.F.H. (1962) *Chemistry of Combustion Reactions*, Butterworths, London.

Mulcahy, M.F.R. (1973) *Gas Kinetics*, Nelson, London.

5.6 PROBLEMS

1. In a cylindrical vessel whose surface destroys hydrogen atoms very efficiently, solution of the diffusion equation leads to the result [98] that the effective first-order 'rate constant' for removal of hydrogen atoms at the wall is

$$f_w = 23D/d^2$$

where d is the diameter of the vessel and D is the diffusion coefficient of hydrogen atoms through the reaction mixture. If, in the hydrogen–oxygen reaction, chain branching is due to

$$H + O_2 \rightarrow OH + O$$

Show that at the first explosion limit

$$kpy = 11.5D/d^2$$

where p is the total pressure and y is the oxygen mole fraction. From this result show that for a fixed temperature at the first limit: (a) d = constant in a particular reaction mixture, and (b) py = constant in a given vessel. Finally, what effect will the addition of inert gas have on the explosion limit pressure?

2. The diffusion coefficient of hydrogen atoms through a hydrogen–oxygen

mixture is given by

$$D = 0.0531\,T^{1.8}/(1 + 0.62y)p$$

where y is the mole fraction of oxygen in the mixture and p is the total pressure in torr. The first explosion limit in a cylindrical vessel, $d = 102$ mm, has been measured by Kurzius and Boudart [101] who obtained the following results:

Mixture						
$2H_2 + O_2$	T(K)	965	931	899	871	847
	p (torr)	2.43	2.72	2.94	3.29	3.64
$9H_2 + O_2$	T(K)	997	959	896	859	833
	p (torr)	4.43	4.76	5.80	6.72	7.67

The vessel was coated with magnesium oxide, a surface on which hydrogen atoms are efficiently destroyed. Use this information, and that in the previous question, to calculate the Arrhenius parameters of the branching reaction.

3. Measurements of the second explosion limit (p) in stoichiometric mixtures of hydrogen and oxygen gave the following results:

T(K)	761	782	800	820
p (torr)	40	56	75	100

If the activation energy of the termolecular terminating step is -7 kJ mol^{-1}, calculate the activation energy of the branching step on the assumption that all HO_2 radicals are destroyed.

4. At a particular temperature, the second explosion limit (p) in mixtures of hydrogen and oxygen varied with composition as follows:

p (torr)	54	66	78	90
p_{O_2}(torr)	20	40	60	80

Use these results to obtain the efficiency of oxygen relative to that of hydrogen in collisions with HO_2, again assuming that all HO_2 radicals are removed at the wall.

5. The kinetic theory of gases leads to the expression for the frequency of collisions between two species A and B in a mixture

$$Z_{A,B} = \left[\frac{8RT(M_A + M_B)}{M_A M_B}\right]^{1/2} \sigma_{AB}^2 n_A n_B$$

where M_A and M_B are the molar masses, n_A and n_B are the numbers of molecules per unit volume, and $\sigma_{AB} = (\sigma_A + \sigma_B)/2$. Calculate the relative frequency of collisions between HO_2 and oxygen, and HO_2 and hydrogen, and compare your answer with the collision efficiency obtained in question 4.

$$\sigma_{O_2} = 0.36 \text{ nm}; \; \sigma_{H_2} = 0.27 \text{ nm}; \; \sigma_{HO_2} = 0.50 \text{ nm}.$$

6

Combustion of hydrocarbons

6.1 CHARACTERISTICS OF HYDROCARBON COMBUSTION

In the simple static systems already described, two regions with different reaction behaviour, which depend primarily on the temperature and pressure and are separated by a very sharp boundary, may be distinguished. On one side of the boundary combustion is slow and may even be unmeasurable, whilst on the other explosive reaction occurs. In the case of hydrocarbons and organic materials generally, the situation is considerably more complex. As before, there is a sharp boundary at which true ignition occurs, characterized by an explosion in a static system, and a yellow flame associated with carbon formation in a flow system. In the 'slow' combustion region, several separate phenomena may be distinguished (Fig. 6.1) and these can be illustrated by referring to the various pressure–time curves observed when a fuel–oxygen mixture reacts in a static system under different initial conditions (Fig. 6.2). In such a system, so long as there are no significant temperature changes, the pressure change is normally a good measure of the extent of reaction and the pressure–time measurements are therefore a useful way of following the reaction. The curves in Fig. 6.2 relate to a fixed initial temperature (T_1 in Fig. 6.1) and gradually increasing initial pressures. At pressures below p_l, slow reaction takes place: this is the region of low-temperature *slow combustion* and the pressure–time curves are sigmoidal (curves a and b in Fig. 6.2) showing an initial slow exponential acceleration due to degenerate chain branching (see later), followed by a decrease in rate as the reactants are consumed. The major product in this region is normally the alkene with the same number of carbon atoms as the fuel, but peroxides, aldehydes, ketones and alcohols are also formed, as are carbon

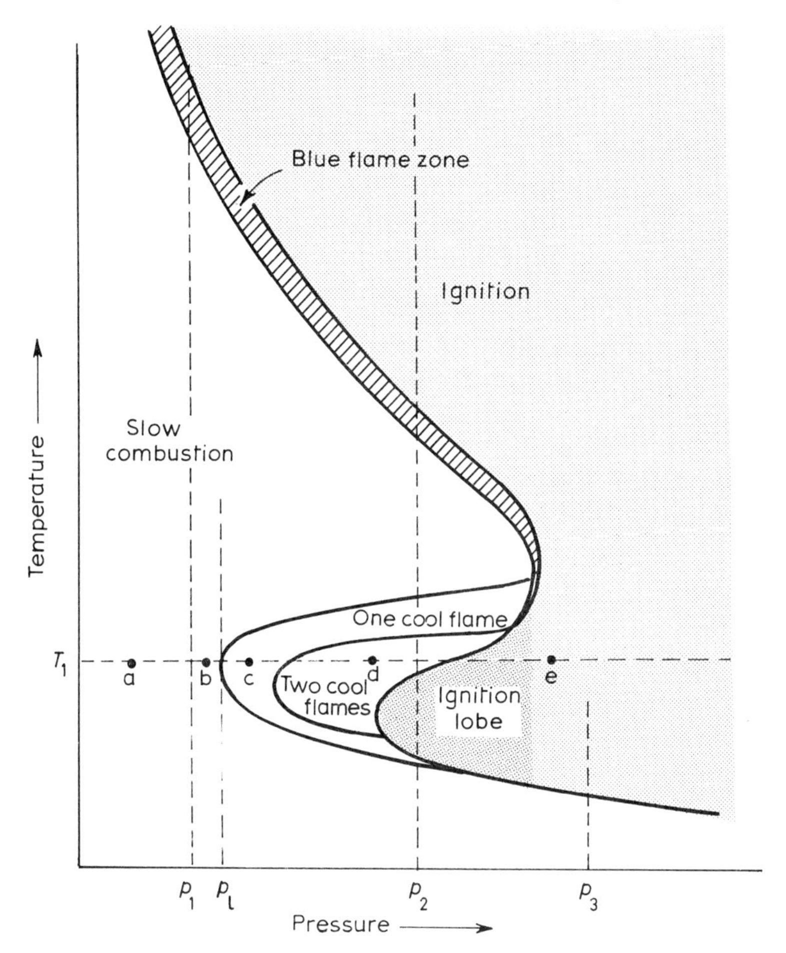

Fig. 6.1 Schematic representation of a typical ignition diagram.

monoxide and water. As the initial reactant pressure is increased, a pale blue flame is seen to traverse the vessel. This is accompanied by a momentary pressure pulse during which the temperature may increase by about 150 K before the system reverts to low-temperature slow combustion (Fig. 6.2, curve c). The duration of a *cool flame* is typically one second. At slightly higher pressures, a succession of these so-called cool flames may be seen (Fig. 6.2, curve d). Further increase in temperature results in *ignition*. Often ignition is preceded by a cool flame and this process is known as *two-stage ignition* (Fig. 6.2, curve e). In the cool flame region, in addition to the products already mentioned,

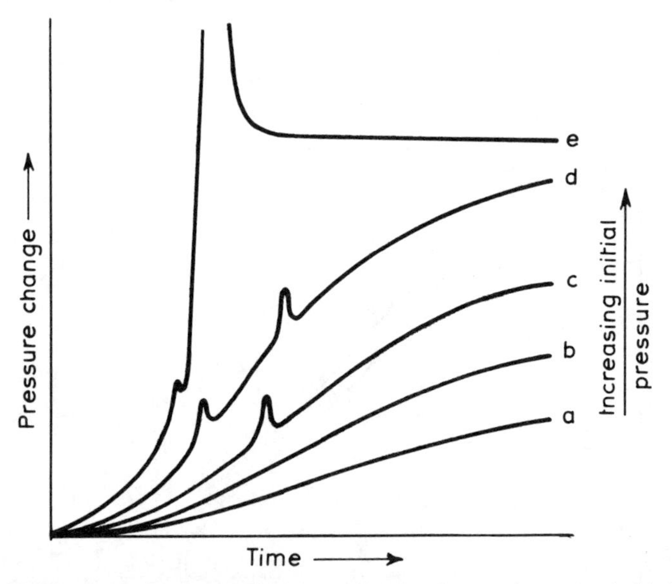

Fig. 6.2 Pressure–time curves during combustion in a static system. (a, b) slow combustion, (c) one cool flame, (d) two cool flames, (e) two-stage ignition. The letters correspond to the points marked in Fig. 6.1.

cyclic ethers are formed in varying amounts while ignition results in a different product spectrum in which lower alkenes and carbon monoxide are prominent.

The boundaries separating the regions in which the various types of behaviour are observed as shown diagrammatically in Fig. 6.1. At both low and high pressures (p_1 and p_3) explosion occurs as the temperature is raised, without intervention of cool flames. Along the line corresponding to an intermediate initial pressure p_2 at first raising the initial reactant temperature causes slow reaction to give way to cool flames and then ignition. However, further increase in temperature results in ignition being replaced by cool flames and then slow reaction before at even higher temperatures, ignition again occurs. This is a manifestation of the so-called *negative temperature coefficient* of the rate which also reveals itself if the rate of reaction of mixtures of constant composition is plotted as a function of temperature (Fig. 6.3) when the region where the rate falls as the temperature is raised can be clearly seen. The form of diagrams such as Fig. 6.3 has led to the subdivision of the slow combustion zone into *high-temperature* and *low-temperature* regions. This high-temperature region of slow combustion must be clearly distinguished from that corresponding to true

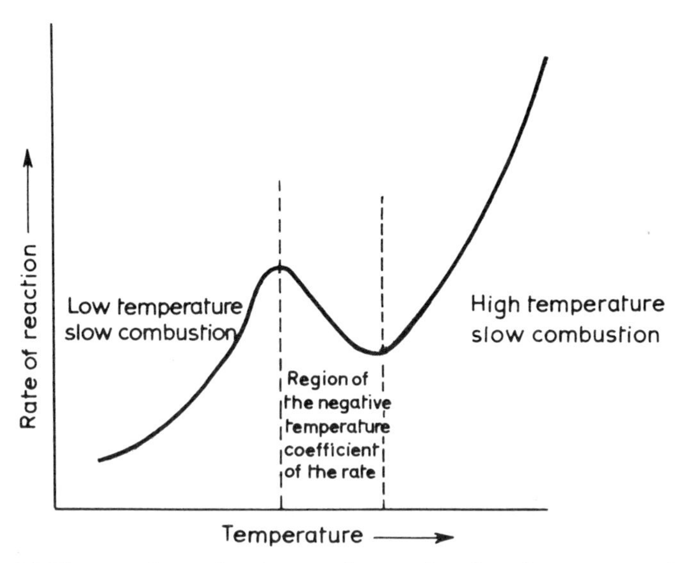

Fig. 6.3 The rate of a combustion reaction as a function of temperature showing the region of negative temperature coefficient of the rate.

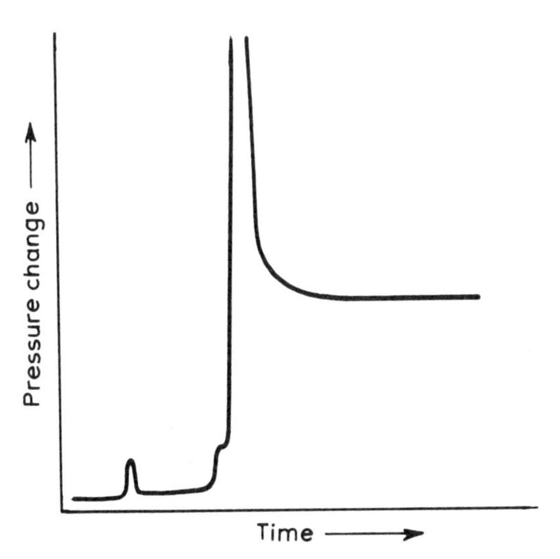

Fig. 6.4 Pressure–time curve showing delayed ignition.

ignition in which temperatures close to the adiabatic flame temperature may be reached.

Other phenomena are sometimes observed. In the region where multiple cool flames are expected, a single cool flame occurs followed, after an appreciable period, by a second cool flame and ignition: this is termed *delayed ignition* (Fig. 6.4). In some systems, the ignition region may be separated from that in which slow combustion occurs by a narrow zone in which an intense blue flame is to be seen.

Several of these phenomena in static systems have their counterparts in flow systems. Cool flames can be stabilized in flow systems and may be seen alone or followed by an intense blue flame. In some circumstances the blue flame may be followed by the yellow flame usually associated with true ignition.

Ignition diagrams such as that shown in Fig. 6.1 have been plotted for many combustion systems. They are generally similar in form and exhibit one or more regions called *lobes* in which ignition requires a higher reactant pressure as the temperature is increased. While the pressure corresponding to p_l depends strongly on the nature of the fuel and the composition of the fuel–oxidant mixture, the temperature range in which cool flames are observed is normally from about 250 to 350° C. A diagram such as Fig. 6.1 is often referred to as describing the *morphology* of the system and it is apparent that combustion in static systems is exceedingly complex. Nevertheless, a great deal of effort has been devoted to correlating the shape of these diagrams, in particular, that of the lobes, with the details of the chemistry of the reacting system [108, 109]. This is a difficult task because the situation is greatly complicated by the temperature and concentration gradients which are set up as the reaction develops and by the inevitable consumption of reactants; surface reactions may also play a part. These factors and others often combine to make ignition diagrams highly specific to the particular system employed and of little general value [110].

The present chapter deals only with the fundamental mechanism and basic chemistry of the elementary steps. Cool flames and other specialized matters involving the interplay of physical and chemical processes will be dealt with in the next chapter.

6.2 DEGENERATE BRANCHING

It is apparent from the description above that the greater chemical complexity of organic materials leads to combustion phenomena which are quite different from those characteristic of simpler systems.

Even where comparable phenomena occur, a significant difference in quantitative behaviour appears. Whilst the simpler systems show short induction times and very rapid growth of reaction rates, the oxidation of a hydrocarbon often displays a long induction period (often lasting for several minutes and occasionally much longer) which is followed by a much slower build-up in reaction rate. Eventually the rate begins to fall due to consumption of reactants before a true explosion can develop. Clearly, although branching-chain reactions must be taking place, they differ from those of simpler systems.

Semenov [111] suggested that in these systems branching is not due to a reaction between a radical and a fuel or oxidant molecule but, instead, a relatively stable intermediate is formed. This intermediate has an appreciable lifetime and is able to react in two ways either to give radicals, hence leading to chain branching, or reverting to stable products, thus

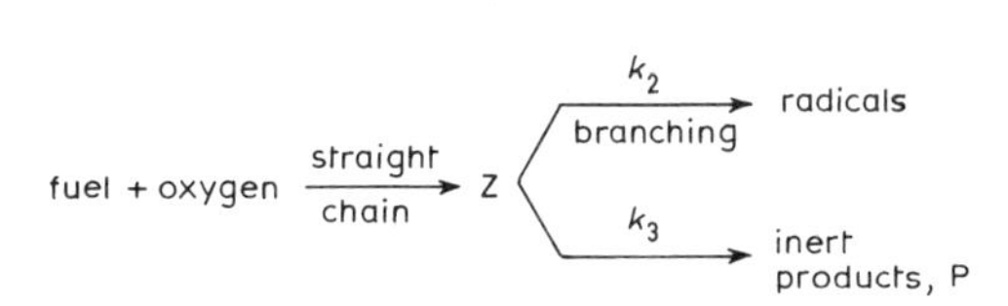

This mechanism is referred to as *delayed* or *degenerate branching* while the overall reaction itself is described as a *degenerate explosion* to distinguish it from a true explosion: the important feature is, of course, that not all species Z lead to branching. An alternative proposal, due to Knox [112–114], suggests that the degeneracy arises in the propagation reaction, so that Z now always leads to branching but is formed only occasionally from the straight-chain cycle.

fuel + oxygen $\xrightarrow[\text{chain}]{\text{straight}}$ Inert products, P

fuel + oxygen $\xrightarrow[\text{chain}]{\text{straight}}$ Z $\xrightarrow{\text{branching}}$ radicals

The actual rate of reaction in the Semenov model, as measured by the appearance of products, is given by

$$\frac{d[P]}{dt} = k_3[Z] \tag{6.1}$$

If it is assumed that each chain cycle produces a single molecule of intermediate, then the rate of formation of Z is given by

$$\frac{d[Z]}{dt} = \bar{\nu}k_1 + \bar{\nu}k_2[Z] - k_2[Z] - k_3[Z] \tag{6.2}$$

where $\bar{\nu}$ is the average number of cycles undergone by a radical before the chain terminates and k_1 is the rate of production of chain centres by some initiation reaction. Integrating twice, assuming that [Z] and [P] are equal to zero when $t = 0$, gives

$$[P] = \frac{\bar{\nu}k_1k_3}{[k_2(\bar{\nu}-1)-k_3]^2}\{\exp[k_2(\bar{\nu}-1)-k_3]t - [k_2(\bar{\nu}-1)-k_3]t - 1\} \tag{6.3}$$

Once the concentration of products becomes appreciable, only the exponential term will be important, and the time dependence of the progress of reaction can be expressed formally by

$$[P] = C' \exp(\phi t) \tag{6.4}$$

where the net branching factor ϕ is equal to $k_2(\bar{\nu}-1) - k_3$.

The formal behaviour predicted by this mechanism is therefore identical to the usual branching-chain explosion (see Chapter 2), i.e. an induction period of the order of $1/[k_2(\bar{\nu}-1)-k_3]$ followed by an exponential increase in rate with a growth constant of $[k_2(\bar{\nu}-1)-k_3]$.

If the alternative mechanism due to Knox applies, then the k_3 term may be neglected and the chain length $\bar{\nu}$ must be multiplied by a factor β which gives the fraction of chain cycles leading to Z. The branching factor now becomes $k_2(\beta\bar{\nu}-1)$. The important features of the Semenov approach are that Z has a long lifetime and that when it reacts it does so to give mostly inert products. On the other hand, in this alternative representation, the intermediate can have a much shorter lifetime and hence lower concentration, whilst still yielding the same branching factor and hence the same kinetics. The validity of the particular mechanism in any given oxidation is therefore intimately bound up with the chemical nature of the intermediate. The nature of species which can yield degenerate branching characteristics is discussed later.

Semenov [115] has also considered the complete course of a degenerate branching reaction and has demonstrated that at low extents of reaction, the rate will show an exponential increase which tails off as the reactants are consumed. Typical curves are shown in Fig. 6.5 in which the S-shaped line with the greater curvature corresponds to a lower rate for the initiation reaction. The

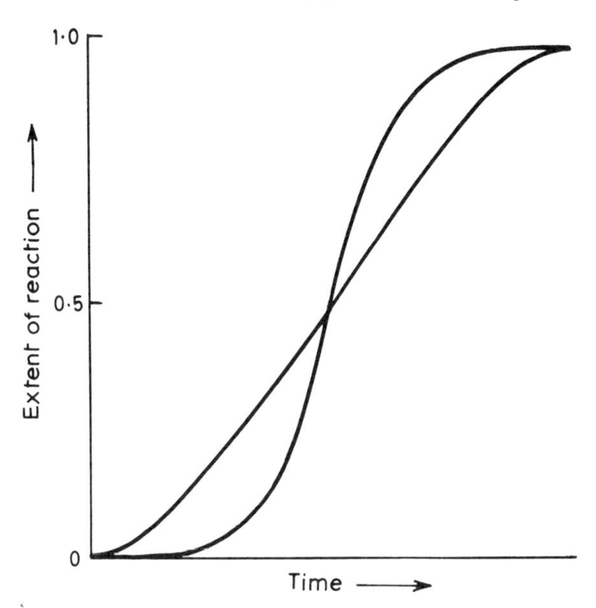

Fig. 6.5 Typical reaction behaviour predicted by degenerate branching theory. The more S-shaped curve is associated with a lower rate for the initiation reaction.

pressure–time curves observed in slow combustion reactions in a static system are frequently of this form (compare Fig. 6.2, curves a and b).

6.3 SUMMARY OF PREDOMINANT REACTIONS

The oxidation of organic compounds involves a wide variety of reactive intermediates and leads to a correspondingly large number of products. Within the scope of this volume it is only possible to give the merest outline of the processes taking place, even though the discussion has been restricted mainly to hydrocarbons because of their commercial importance. Before dealing with specific examples, the reactions which occur may conveniently be classified according to the chemical nature of the intermediate and the type of reaction in which it participates.

6.3.1 Initiation

At high temperatures, the primary step is almost certainly the decomposition of a hydrocarbon molecule to give, amongst other species, an alkyl radical. At low temperatures, direct decomposition is too slow and the primary step involves both fuel and oxygen. For a saturated hydrocarbon, this will almost certainly be a hydrogen abstraction

$$RH + O_2 \rightarrow R + HO_2 \tag{6.5}$$

although surface initiation may also play a part.

With unsaturated hydrocarbons, reactions analogous to Reaction 6.5 still occur, especially when the radical formed is a very stable one, for example, allyl, but addition of oxygen to the double bond is also important [116, 117]. The initial adduct will be unstable and will decompose rapidly to form a range of intermediates amongst which aldehydes are likely to be the most important

$$\begin{array}{c} RCH{=}CHR' + O_2 \rightarrow \begin{array}{c} RCH{-}CHR' \\ | \quad\;\; | \\ O{-}O \end{array} \\ \downarrow \\ RCHO + R'CHO \end{array} \tag{6.6}$$

6.3.2 Reactions of R Radicals

Reaction 6.5 results in the production of alkyl radicals, R. These radicals, which are also formed in other processes (see below), react in one of three ways:

(a) *Decomposition.* All but the simplest alkyl radicals decompose to give an alkene and another alkyl radical, for example,

$$RCH_2CHR' \rightarrow R + CH_2 = CHR' \tag{6.7}$$

Clearly, if the reaction is rapid, the oxidation of a saturated hydrocarbon may quickly transform to the oxidation of a lower alkene.

(b) *Reaction with oxygen.* This process plays a most important part in hydrocarbon oxidation. Although there has been much debate, it now seems to be established that the direct bimolecular reaction

$$R + O_2 \rightarrow \text{Alkene} + HO_2 \tag{6.8}$$

does take place (except with the methyl radical when alkene formation is not possible) and at least for simple alkanes at temperatures up to 500° C this reaction is the most important product-forming step.

Alkyl radicals also react with oxygen in a fast reversible exothermic process to form alkylperoxy radicals

$$R + O_2 \rightleftharpoons RO_2 \tag{6.9}$$

Table 6.1 Estimated values of K for the equilibrium $R + O_2 \rightleftharpoons RO_2$ [93].

R	$\log_{10}K(\text{atm}^{-1})$		
	300 K	500 K	700 K
CH_3	12.53	4.78	1.47
C_2H_5	12.78	4.74	1.30
i-C_3H_7	12.73	4.84	1.46
t-C_4H_9	12.98	4.60	0.97

The equilibrium constant of Reaction 6.9 has been reliably estimated for a large number of alkyl radicals using thermochemical data. A few values are listed in Table 6.1 and it can be seen that under conditions typical of slow combustion the equilibrium value of $[RO_2]/[R]$ is likely to be greater than unity. Alkylperoxy radicals are therefore important intermediates in the formation of many combustion products (see below).

(c) *Isomerization.* This may compete with the alternative reactions of alkyl radicals thereby affecting the proportions of the various isomeric species present. Such reactions are important for the more complex hydrocarbons and for alkenes.

The regeneration of alkyl radicals after their initial production in Reaction 6.5 will take place either by abstraction of hydrogen from the parent hydrocarbon by other radicals in the system, for example,

$$OH + RH \rightarrow R + H_2O \tag{6.10}$$

or by decomposition of less stable intermediates present, for example,

$$RCO \rightarrow R + CO \tag{6.11}$$

6.3.3 Reactions of RO_2 radicals

At least at low temperatures, many of the products of combustion are formed via the RO_2 radical which can undergo a wide range of reactions.

(a) *Decomposition.* Particularly when RO_2 contains excess energy carried over from its formation it may decompose back to R and O_2.

(b) *Isomerization.* This usually occurs via internal hydrogen abstraction and the resulting hydroperoxyalkyl radical QOOH can then undergo various reactions including loss of OH followed by cyclization and cleavage of a carbon–carbon bond together with the elimination of an OH or HO_2 radical.

$$RO_2 \rightarrow QOOH \rightarrow \text{Molecular product} + OH(\text{or } HO_2) \quad (6.12)$$

In any but the simplest alkyl radicals, several internal hydrogen abstractions may be possible and their relative importance will depend on the steric and energetic factors involved.

Starting radical: $R(H)C(OO)CH_2CH_2CH_2R'$

$\rightarrow R(OOH)C\dot{}CH_2CH_2CH_2R'$ — 1:3 H transfer, 4-membered ring transition state (6.13)

$\rightarrow R(H)(OOH)C\,\dot{C}HCH_2CH_2R'$ — 1:4 H transfer, 5-membered ring transition state (6.14)

$\rightarrow R(H)(OOH)C\,CH_2\dot{C}HCH_2R'$ — 1:5 H transfer, 6-membered ring transition state (6.15)

$\rightarrow R(H)(OOH)C\,CH_2CH_2\dot{C}HR'$ — 1:6 H transfer, 7-membered ring transition state (6.16)

A small highly strained ring in the transition state implies a large activation energy. The strength of the C–H bond from which internal abstraction occurs is also important, and it is easier to break a tertiary C–H bond than a secondary one which in turn is more easily broken than a primary one. The pre-exponential factors and activation energies for various internal hydrogen abstraction reactions are given in Table 6.2 and these figures enable an assessment to be made of the likely importance of possible internal H-abstractions in any given RO_2 radical.

(c) *Intermolecular abstraction.* Provided the RO_2 radicals have an appreciable lifetime, they can also participate in external hydrogen abstraction reactions in the same way as other radicals present.

$$RO_2 + RH \rightarrow ROOH + R \quad (6.17)$$

The important feature of this reaction is the production of a molecular species, the alkyl hydroperoxide, ROOH, which can lead to degenerate branching.

(d) *Reaction with other radicals.* If reaction rates are high, radical concentrations may be appreciable and then radical–radical processes

Table 6.2 Arrhenius parameters for internal H atom transfer reactions in RO_2 radicals, namely, $RO_2 \rightarrow QOOH$ [118].

Type	$A(s^{-1})$ (per C–H)	$E(kJ\,mol^{-1})$
1,4 p	8×10^{12}	145
1,5 p	1×10^{12}	117
1,6 p	1.25×10^{11}	98
1,7 p	1.55×10^{10}	82
1,3 s	6.4×10^{13}	168
1,4 s	8×10^{12}	125
1,5 s	1×10^{12}	100
1,6 s	1.25×10^{11}	80
1,7 s	1.55×10^{10}	64
1,3 t	6.4×10^{13}	153
1,4 t	8×10^{12}	110
1,5 t	1.0×10^{12}	84
1,6 t	1.25×10^{11}	64
1,7 t	1.55×10^{10}	48

For an explanation of the basis of these data see [118].

which have low activation energies may become important in chain propagation and product formation [119].

For alkylperoxy radicals, the most likely reactions are with other RO_2 radicals, including HO_2

$$RO_2 + R'O_2 \rightarrow RO + R'O + O_2 \tag{6.18}$$

$$RO_2 + HO_2 \rightarrow RO + OH + O_2 \tag{6.19a}$$

$$RO_2 + HO_2 \rightarrow ROOH + O_2 \tag{6.19b}$$

6.3.4 Reactions of QOOH radicals

These radicals react in various ways, all of which result in chain propagation.

(a) *Cyclization to O-heterocycles, with elimination of OH.* Fission of the O–O bond may be followed by ring closure to give a cyclic ether. For example, the hydroperoxyalkyl species formed in Reaction 6.15 may give an oxetan by

$$\mathrm{R(H)C(OOH)CH_2CHCH_2R'} \longrightarrow \mathrm{R(H)C(O)CH_2CHCH_2R'} + \mathrm{OH} \longrightarrow \mathrm{R(H)C\langle O,CH_2 \rangle CHCH_2R'} \tag{6.20}$$

The analogous reaction of the species formed in Reaction 6.14 gives an oxiran, while a furan results from the species formed in Reaction 6.16.

(b) *Decomposition.* There are several modes of decomposition of QOOH available depending on its structure. Considering as an example again the species formed in Reaction 6.15, this may decompose to give an aldehyde and an alkene

$$\mathrm{R(H)C(OOH)CH_2CHCH_2R'} \longrightarrow \mathrm{R(H)C{=}O + OH + CH_2{=}CHCH_2R'} \tag{6.21}$$

while the radical formed in Reaction 6.13 gives a ketone

$$\mathrm{RC(OOH)CH_2CH_2CH_2R'} \longrightarrow \mathrm{RC({=}O)CH_2CH_2CH_2R' + OH} \tag{6.22}$$

(c) *Reaction with oxygen.* A hydroperoxyalkyl radical can behave as a simple alkyl radical and add oxygen giving eventually a dihydroperoxyalkyl radical, for example,

$$\underset{\mathrm{OO}}{\mathrm{RCH}}(\mathrm{CH_2})_3\mathrm{R'} \rightarrow \underset{\mathrm{OOH}}{\mathrm{RCH}}(\mathrm{CH_2})_2\mathrm{CHR'} \xrightarrow{+\mathrm{O_2}} \underset{\mathrm{OOH}}{\mathrm{RCH}}(\mathrm{CH_2})_2\underset{\mathrm{OO}}{\mathrm{CHR'}} \tag{6.23}$$

It appears that this reaction is most important when the QOOH radicals are formed by 1:5 H transfer [94].

This dihydroperoxy radical will react further and may decompose into products such as aldehydes and ketones (and an OH) or, particularly at lower temperatures, abstract hydrogen from a suitable donor giving a dihydroperoxide molecule.

6.3.5 Reactions of peroxides and hydroperoxides

The O–O bond in these molecules is a weak one and the only significant reaction of peroxides and hydroperoxides is decomposition: the reaction is, however, of supreme importance because it normally yields two radical species. These molecules can therefore behave as degenerate branching intermediates.

$$\mathrm{ROOR'} \rightarrow \mathrm{RO + OR'} \tag{6.24}$$

$$\mathrm{ROOH} \rightarrow \mathrm{RO + OH} \tag{6.25}$$

6.3.6 Reactions of alkoxy radicals

Alkoxy radicals, RO, are formed in the decomposition of hydroperoxides and in radical–radical reactions involving RO_2. Like alkyl radicals they may decompose, react with oxygen or abstract hydrogen.

(a) *Decomposition.* Depending on the nature of the RO radical, several modes of decomposition may occur but the primary fission always appears to take place at the carbon atom adjacent to the oxygen, leading to an aldehyde and an alkyl radical, for example,

$$R(R')C(H)O \longrightarrow R + R'CHO \quad (6.26)$$

This provides an alternative route for the formation of lower aldehydes in combustion systems.

(b) *Reaction with oxygen.* Reaction with oxygen can lead to HO_2 radicals and aldehydes, thus

$$RCH_2O + O_2 \rightarrow RCHO + HO_2 \quad (6.27)$$

Secondary alkoxy radicals yield ketones by a similar process.

(c) *Abstraction.* RO radicals will abstract hydrogen in exactly the same way as other radicals, in this case producing alcohols and regenerating R′ radicals.

$$RO + R'H \rightarrow ROH + R' \quad (6.28)$$

The importance of this reaction is that it leads to the formation of alcohols.

6.3.7 Reactions of aldehydes

Two routes to the formation of aldehydes in combustion systems have already been described. Because aldehydes possess a labile hydrogen atom, the major process removing them almost certainly leads to RCO radicals

$$RCHO \rightarrow RCO + H \quad (6.29)$$

either by direct decomposition as above, or more probably by hydrogen abstraction, including reaction with oxygen.

6.3.8 Reactions of RCO radicals

RCO radicals can react in two ways. At higher temperatures decomposition, Reaction 6.11, will occur giving R radicals and carbon monoxide. At lower temperatures and/or higher pressures, reaction with oxygen will also be important

$$RCO + O_2 \rightarrow RCO_3 \tag{6.30}$$

6.3.9 Reactions of RCO_3 radicals

Apart from decomposition to the initial reactants, which is not mechanistically significant, the major reaction is hydrogen abstraction to give the peracid and an R′ radical

$$RCO_3 + R'H \rightarrow RCO_3H + R' \tag{6.31}$$

6.3.10 Reactions of peracids

Since peracids contain peroxide linkages, the major reaction will be fission at the O–O bond, to give two radicals

$$RCO_3H \rightarrow RCO_2 + OH \tag{6.32}$$

This intermediate is therefore also responsible for branching although since it is derived from the aldehyde, it may prove equally viable to consider only reactions of the latter.

The RCO_2 radical is a source of the carbon dioxide

$$RCO_2 \rightarrow R + CO_2 \tag{6.33}$$

which is a product of low-temperature combustion.

The various reactions listed above are summarized schematically in Fig. 6.6.

6.3.11 Degenerate branching intermediates

From the brief summary of reactions above, it can be seen that aldehydes, peroxides and hydroperoxides, and peracids may all behave as degenerate-branching intermediates. Arguments continue as to the nature of the intermediates in any given system but it appears likely that hydrogen peroxide or formaldehyde is the most important intermediate for simple fuels in the ‘high-temperature’ region. In the ‘low-temperature’ region, higher aldehydes may be important but it is probable that the presence of hydroperoxides and to a lesser extent, peracids dominates the course of reaction.

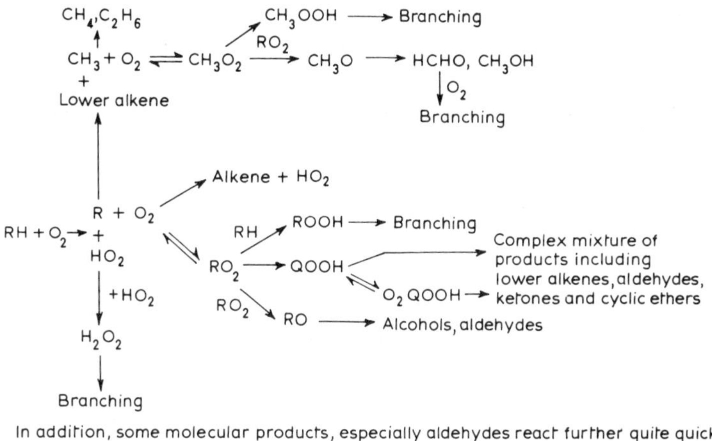

Fig. 6.6 Outline reaction scheme for hydrocarbon combustion.

6.3.12 Reactions of OH and HO_2 radicals

So far we have considered the reactions of species derived from the fuel. The main oxidizing species are the oxygen molecules themselves, oxygen atoms, hydroxyl and hydroperoxy radicals, of which the last two are the most important particularly at low temperatures.

Oxygen atoms become important only at high temperatures. Their reactions are not well understood although as di-radical species they are often responsible for chain branching, for example,

$$O + H_2 \rightarrow OH + H \qquad (6.34)$$

The main reactions of OH and HO_2 radicals are hydrogen abstraction processes such as Reaction 6.10 although under some circumstances HO_2 undergoes radical–radical reactions such as Reactions 6.19a and 6.19b.

6.4 CHAIN CYCLES IN SLOW COMBUSTION

Although the number of possible reactions in any oxidation system is very great, we are primarily interested in the rapid chain reactions which are responsible for the major consumption of reactants, rather than in the numerous minor processes which give rise to a variety of side products. It is possible to distinguish one or two chain cycles which may dominate a phase of reaction under an appropriate set of conditions.

In discussing these it is important to recognize that OH is more reactive and less selective in its attack on alkanes compared with HO_2 and RO_2; the Arrhenius parameters for these reactions have been given in Table 5.3.

6.4.1 Peroxy radical chains

The sequence

$$R + O_2 \rightleftharpoons RO_2 \tag{6.9}$$

$$RO_2 + RH \rightarrow ROOH + R \tag{6.17}$$

forms a simple example of a chain cycle which is frequently important at lower temperatures. It also leads to the formation of ROOH which is believed to be an important degenerate chain-branching intermediate and causes the stationary concentration of R to increase

$$ROOH \rightarrow RO + OH \tag{6.25}$$

$$RO + RH \rightarrow ROH + R \tag{6.28}$$

$$OH + RH \rightarrow H_2O + R \tag{6.10}$$

6.4.2 Hydroperoxy radical chains

An alternative, which is likely to become more important at higher temperatures, is

$$R + O_2 \rightarrow \text{Alkene} + HO_2 \tag{6.8}$$

$$RH + HO_2 \rightarrow R + H_2O_2 \tag{6.35}$$

This sequence produces alkene and hydrogen peroxide in stoichiometric amounts. Decomposition of hydrogen peroxide would lead to degenerate branching in this mechanism at temperatures above 450° C.

In the early stages of the oxidation of several simple alkanes at temperatures around 400° C, most of the fuel consumed is converted to the conjugate alkene [120], but seldom is hydrogen peroxide detected in such large amounts. In addition, the rate of attack by HO_2 on the alkane is insufficient to sustain the observed rate of reaction. As the reaction proceeds it appears that HO_2 radicals are converted to the more reactive OH by means which are not fully understood, although the reaction

$$HO_2 + RO_2 \rightarrow OH + RO + O_2 \tag{6.19a}$$

may play some part, together with reaction of HO_2 with alkenes to give oxirans and OH. The net effect is that the hydroperoxy chain is transformed into one propagated by OH radicals.

6.4.3 Hydroxyl radical chains

Hydroxyl radicals are highly reactive and will readily abstract hydrogen atoms from alkanes giving water which is, of course, an important product of combustion. Many of the reactions leading to stable products, for example, Reactions 6.20–6.22, also result in formation of an OH radical, and a further source is Reaction 6.19a. A chain cycle propagated by OH radicals is established

$$RH + OH \rightarrow R + H_2O \qquad (6.10)$$

$$R + O_2 \rightarrow \text{Product} + OH \qquad (6.36)$$

Many of the basic features of hydrocarbon oxidation can be explained in terms of these chain cycles, the relative importance of each depending markedly on the conditions such as temperature and concentration under which the reaction is taking place.

6.5 THE OXIDATION OF HYDROCARBONS

6.5.1 The oxidation of methane

Because methane is the simplest hydrocarbon, its oxidation has been extensively studied but even with this molecule the processes are not fully understood [121]. Below 450° C, reaction is very slow, but above this temperature gas-phase combustion takes place with increasing speed.

Initiation presumably occurs either by some surface process yielding radicals or in the gas phase by

$$CH_4 + O_2 \rightarrow CH_3 + HO_2 \qquad (6.37)$$

The HO_2 radicals may either attack methane

$$CH_4 + HO_2 \rightarrow CH_3 + H_2O_2 \qquad (6.38)$$

or undergo radical–radical reactions such as

$$HO_2 + HO_2 \rightarrow H_2O_2 + O_2 \qquad (6.39)$$

$$CH_3O_2 + HO_2 \rightarrow CH_3OOH + O_2 \qquad (6.40)$$

Clearly the dominant alkyl radical is methyl and its reaction with oxygen plays an important part in the overall process, as it also does in the combustion of other alkanes. In spite of the apparent simplicity, and many investigations, the precise path of this process is still not clearly established. The main products of the methyl + oxygen reaction are methanol and formaldehyde, the proportions depending on the conditions. At relatively low temperatures there is no doubt that the methylperoxy radical is formed

$$CH_3 + O_2(+M) \rightleftharpoons CH_3O_2(+M) \qquad (6.41)$$

and in most combustion systems this will react with other radicals, particularly CH_3, CH_3O_2 and HO_2, when these are present in sufficient concentration

$$CH_3O_2 + CH_3 \rightarrow 2CH_3O \qquad (6.42)$$

$$CH_3O_2 + CH_3O_2 \rightarrow 2CH_3O + O_2 \qquad (6.43)$$

$$CH_3O_2 + HO_2 \rightarrow CH_3OOH + O_2 \qquad (6.40)$$

$$CH_3O_2 + HO_2 \rightarrow CH_3O + OH + O_2 \qquad (6.44)$$

Further reactions of the methoxy radical will give methanol and formaldehyde

$$CH_3O + RH \rightarrow CH_3OH + R \qquad (6.45)$$

$$CH_3O + M \rightarrow HCHO + H + M \qquad (6.46)$$

$$CH_3O + O_2 \rightarrow HCHO + HO_2 \qquad (6.47)$$

At higher temperatures, there is some evidence for the reaction

$$CH_3 + O_2 \rightarrow HCHO + OH \qquad (6.48)$$

This is not a simple bimolecular process but probably involves the methylperoxy radical which undergoes an intramolecular hydrogen atom transfer

$$CH_3O_2 \rightarrow CH_2OOH \qquad (6.49)$$

followed by fission of the O–O bond

$$CH_2OOH \rightarrow HCHO + OH \qquad (6.50)$$

Since the hydrogen transfer involves breaking a primary C–H bond and a four-membered transition state ring this path has a high energy barrier and will only become important at high temperatures.

The hydroxyl radicals produced in Reaction 6.48 will propagate a chain

$$OH + CH_4 \rightarrow H_2O + CH_3 \qquad (6.51)$$

in which water and formaldehyde are the molecular products.

Formaldehyde leads to degenerate chain branching by

$$HCHO + O_2 \rightarrow CHO + HO_2 \tag{6.52}$$

Branching may also occur by

$$H_2O_2 \rightarrow 2OH \tag{6.53}$$

Formaldehyde will also be attacked by OH and HO_2 radicals

$$OH + HCHO \rightarrow H_2O + CHO \tag{6.54}$$

$$HO_2 + HCHO \rightarrow H_2O_2 + CHO \tag{6.55}$$

Formyl radicals, CHO, react with oxygen to give carbon monoxide

$$CHO + O_2 \rightarrow CO + HO_2 \tag{6.56}$$

which can also result from the direct decomposition of the radical

$$CHO + M \rightarrow CO + H + M \tag{6.57}$$

In slow oxidation, reaction chains will terminate at the walls probably by removal of species such as HO_2, although the evidence is not conclusive on this point. The products HCHO, H_2O_2, H_2O and CO predicted by this mechanism can all be detected in the system.

Other products almost certainly arise from alternative reactions of CH_3O_2 radicals. Although these radicals do not abstract hydrogen from the fuel directly, they can do so from other species, for example,

$$CH_3O_2 + HCHO \rightarrow CH_3OOH + CHO \tag{6.58}$$

The methyl hydroperoxide can break down in a branching step to give CH_3O radicals which then abstract further to give methanol.

Because the characteristic oxidation behaviour of organic materials depends on the formation of a degenerate-branching intermediate, and formaldehyde, methyl hydroperoxide and hydrogen peroxide could all perform that function in the present system, it is worth while summarizing the evidence [122] supporting formaldehyde in this role:

(a) The maximum oxidation rate coincides with a maximum in the HCHO concentration.
(b) Addition of formaldehyde reduces or eliminates the induction period but leaves the maximum rate of oxidation unaffected unless an amount in excess of that formed in the reaction is introduced, in which case an initial high rate of oxidation is observed which falls rapidly to the 'normal' rate.

(c) Photolysis of the reaction mixture at wavelengths known to cause decomposition of formaldehyde reduces the induction period and increases the maximum rate.

In the oxidation of methane, and also several other fuels, the maximum formaldehyde concentration is proportional to the fuel concentration but independent of oxygen. This suggests a formal mechanism

$$\mathrm{X + Fuel \xrightarrow{k_1} HCHO + ?} \quad (6.59)$$

$$\mathrm{X + HCHO \xrightarrow{k_2} ?} \quad (6.60)$$

so that

$$[\mathrm{HCHO}] = (k_1/k_2)[\mathrm{Fuel}] \quad (6.61)$$

In the case of methane, the part of Reaction 6.59 is played by Reactions 6.38 and 6.51 followed by Reaction 6.48, while Reaction 6.60 is represented by Reactions 6.54 and 6.55.

The complete reaction scheme (Fig. 6.7) thus involves an initial production of radicals by Reaction 6.37 which results in a primary chain propagated by CH_3 and OH radicals. One of the products of this chain, formaldehyde, augments the supply of radicals by degenerate chain branching but is also removed by a non-branching chain leading to carbon monoxide. Hydroperoxy radicals are formed in various ways and these produce hydrogen peroxide which may function as an additional chain-branching agent. Chains are terminated by removal of propagating species at the walls.

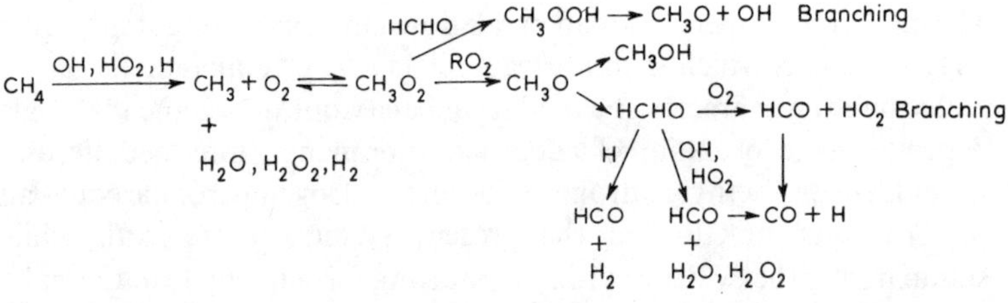

$CH_4 \xrightarrow{OH,\,HO_2,\,H} CH_3 + O_2 \rightleftharpoons CH_3O_2$ (+ H_2O, H_2O_2, H_2)

$CH_3O_2 \xrightarrow{HCHO} CH_3OOH \rightarrow CH_3O + OH$ Branching

$CH_3O_2 \xrightarrow{RO_2} CH_3O$; $CH_3O \rightarrow CH_3OH$; $CH_3O \rightarrow HCHO$

$HCHO \xrightarrow{O_2} HCO + HO_2$ Branching; $HCO \rightarrow CO + H$

$HCHO \xrightarrow{H} HCO + H_2$

$HCHO \xrightarrow{OH,\,HO_2} HCO + H_2O, H_2O_2$; $HCO \rightarrow CO + H$

Fig. 6.7 Outline reaction scheme for slow combustion of methane.

6.5.2 The oxidation of higher hydrocarbons

The intricate morphology of combustion systems has already been described; in addition, complex mixtures of reaction products are formed whose composition depends markedly on the conditions under which reaction takes place. As the molecular weight of the hydrocarbon increases, the number of compounds found in the products rises and, for example, in the combustion of 3-methylpentane in the cool flame region at least 86 distinct products have been detected although, of course, some of these are present in only very small amounts [123].

In many cases, unravelling the situation is made more difficult because some of the primary products (e.g. alkenes) are themselves further oxidized and many of the compounds detected after more than a few per cent conversion of the initial reactants are secondary or even tertiary products.

At temperatures below 200° C, reaction is normally slow unless the chains are initiated by some means such as the addition of a compound which decomposes very rapidly to give radicals. When reaction does occur, then the products are largely the alkyl hydroperoxide

$$R + O_2 \rightleftharpoons RO_2 \qquad (6.9)$$

$$RO_2 + RH \rightarrow ROOH + R \qquad (6.17)$$

Branching is brought about by decomposition of the hydroperoxide

$$ROOH \rightarrow RO + OH \qquad (6.25)$$

both RO and OH being able to propagate the chain,

$$RO + RH \rightarrow R + ROH \qquad (6.28)$$

$$OH + RH \rightarrow H_2O + R \qquad (6.10)$$

At somewhat higher temperatures the equilibrium in Reaction 6.9 is displaced to the left and, furthermore, abstraction by those RO_2 radicals which are formed becomes less important than other reactions of RO_2, for example,

$$RO_2 \rightarrow QOOH \rightarrow \text{products} \qquad (6.62)$$

The displacement of the equilibrium also results in an increasing concentration of R radicals and thus alkene production by

$$R + O_2 \rightarrow \text{Alkene} + HO_2 \qquad (6.8)$$

assumes a greater importance. Indeed, for the C_2–C_4 hydrocarbons, the yields of alkene by this route between 300 and 500° C are very high in the early stages of the reaction.

The products of Reaction 6.62 usually include higher aldehydes, in addition to formaldehyde which is relatively stable at these temperatures and does not participate in branching below 400° C. Other aldehydes may cause some degenerate branching at lower temperatures by, for example,

$$CH_3CHO + O_2 \rightarrow CH_3CO + HO_2 \\ \searrow CH_3 + CO \qquad (6.63)$$

Thus, the formation of very large amounts of degenerate branching agent (hydroperoxide) in the main chain steps gives way to reactions which do not produce branching agent; the latter is now produced in a minor pathway which becomes of decreasing importance as the temperature is raised even more. Consequently, the overall rate of reaction falls; this is the negative temperature coefficient of the rate referred to previously.

As the temperature is raised further, formaldehyde can cause branching by

$$HCHO + O_2 \rightarrow HCO + HO_2 \qquad (6.52)$$

In addition, some hydrogen peroxide is formed by various routes, for example,

$$HO_2 + RH \rightarrow H_2O_2 + R \qquad (6.35)$$

$$HO_2 + HO_2 \rightarrow H_2O_2 + O_2 \qquad (6.39)$$

and this too can bring about chain branching by the gas-phase reaction

$$H_2O_2\ (+M) \rightarrow 2OH\ (+M) \qquad (6.53)$$

which becomes rapid at temperatures above about 400° C. The growing importance of these reactions causes the rate to increase again with temperature. At yet higher temperatures, the decomposition of alkyl radicals to smaller fragments by Reaction 6.7 begins to dominate the other reactions of this species. For example, propyl radicals break down quite rapidly

$$n\text{-}C_3H_7 \rightarrow CH_3 + C_2H_4 \qquad (6.64)$$

Other alkyl radicals react in an analogous fashion and so lower alkenes are generated. At very high temperatures and low oxygen concentrations the reaction becomes almost an oxygen-induced pyrolysis [124].

The customary distinction between low-temperature and high-temperature mechanisms is thus seen to be somewhat artificial. Certainly there are differences in the products of reaction above and below the cool flame region but many of the same reactions are involved in both.

The main product-producing chain can be represented formally by

$$
\begin{array}{ccccccc}
R + O_2 & \rightleftharpoons & RO_2 & \rightleftharpoons & QOOH & \rightleftharpoons & O_2QOOH \\
\downarrow & & \downarrow & & \downarrow & & \downarrow \\
\text{Products} & & \text{Products} & & \text{Products} & & \text{Products}
\end{array}
$$

The rate of formation of a given product can then be written in terms of the rate constants of the various steps.

A crucial factor in determining the composition of the reaction products is the balance between R, O_2 and RO_2 in the equilibrium

$$R + O_2 \rightleftharpoons RO_2 \tag{6.9}$$

Both the forward and backward reactions are very fast, and so Reaction 6.9 is usually assumed to be effectively equilibrated [93], and under normal non-flame combustion conditions the concentration of RO_2 relative to that of R will be appreciable (see Table 6.1).

At high oxygen pressures, or when the reaction rate is high, the concentration of RO_2 radicals may rise considerably and then radical–radical reactions may become major chain-propagating steps and the nature of the products will change for this reason.

In principle, it should be possible to use the information in Tables 5.3, 6.1 and 6.2 to predict the product distribution in cool-flame and slow combustion experiments. However, the conditions are generally so poorly defined, the role of secondary products is so uncertain and, for all but very simple alkanes, the mechanisms are so complex that complete quantitative verification has not been achieved. However, in some cases it has been possible to account satisfactorily for the yields of products in the early stages of reaction [120].

As more knowledge is gained of the behaviour of the primary oxidation products, for example, alkenes, in the reacting mixture it should be possible to reach a much better understanding of the complete combustion process from fuel to CO, CO_2 and water.

6.5.3 High-temperature (flame) combustion of hydrocarbons

In pre-mixed flames, the mechanism of combustion differs markedly from that responsible for low-temperature oxidation. The region here is that corresponding to true ignition in Fig. 6.1. When true ignition occurs the majority of the enthalpy of reaction is released rapidly in a narrow reaction zone leading to the production of very high temperatures. These high temperatures produce steep temperature gradients and the transport of heat or of active centres makes the flame self-propagating. The resulting combustion wave can be stabilized on a burner in the normal way.

The high temperature generated provides the key to the change in mechanism. At high temperature, equilibrium considerations favour dissociation and typical flames in oxygen at 2000–3000 K may contain 10–50% of the fuel in the form of radicals. Furthermore, because the kinetic order of dissociation is lower than that of recombination, radical concentrations will tend to 'overshoot', that is, to exceed the local equilibrium values and will return to them relatively slowly. Although as a result of the high temperature all reactions will be correspondingly faster, because of the increased radical concentrations radical–radical reactions become progressively more significant. The nature of the dominant active intermediates is also likely to change, degenerate-branching species such as peroxides and aldehydes being replaced by simple radicals and atoms. In particular, the reaction of hydrogen atoms becomes very important, due to the increased significance of processes such as Reactions 6.29, 6.34 and

$$RCH_2CH_2 \rightarrow RCH{=}CH_2 + H \qquad (6.7)$$

This change in the nature of the intermediates tends to make chain branching more rapid, the major sequence usually being closely related to that which occurs in the hydrogen–oxygen reaction

$$H + O_2 \rightarrow OH + O \qquad (6.65)$$

$$O + H_2(RH) \rightarrow OH + H(R) \qquad (6.66)$$

$$OH + H_2(RH) \rightarrow H_2O + H(R) \qquad (6.67)$$

The effects of the high temperature once established are self-perpetuating and a transition from slow combustion to true ignition can be effected, under appropriate conditions, simply by supplying a sufficiently active source of ignition, that is, a region of very high temperature or very high radical concentration.

Although the majority of the reaction occurs within a very narrow spatial region, three distinct zones – a pre-heat zone, a true reaction zone, and a recombination zone – may be distinguished. The physical characteristics of these regions have been described in Chapter 3; the chemical processes taking place in them are also quite different.

The nature of the reactions in the pre-heat zone depends very much on the fuel involved. For a very stable molecule like methane, with a first-order decomposition rate constant of only $10^4\ s^{-1}$ even at 2000 K, little or no pyrolysis can occur within the short residence time in the flame. With the majority of hydrocarbons, considerable degradation occurs and the fuel fragments leaving this zone will comprise mainly lower hydrocarbons, alkenes and hydrogen. One major effect of this is that the composition in the reaction zone proper is always very similar, irrespective of the nature of the fuel, thus explaining why flame temperature and burning velocities vary by only a small amount for a wide range of fuels. In the pre-heat zone, the oxygen plays only a catalytic role and is itself little consumed.

The processes occurring in the reaction zone are mainly similar to those already described earlier in this chapter except that their relative importance is considerably altered by the factors discussed above. The dominant reactions are

$$H + O_2 \rightarrow OH + O \qquad (6.65)$$

$$CO + OH \rightarrow CO_2 + H \qquad (6.68)$$

and the velocities of alkane flames can be modelled within a factor of two using the pure H_2–O_2–CO mechanism with the addition of attack by H, O and OH on the hydrocarbon: because of the high temperatures involved the resulting alkyl radicals can be assumed to form CO and H_2O directly at a rate which is effectively infinitely fast. This model is, of course, greatly simplified and a better account of the detailed behaviour of the flames of higher alkanes is given by a mechanism in which initial attack on the fuel by H, O and OH produces alkyl radicals. At the temperatures prevailing in the flame, these break down very rapidly to alkenes and a simple radical, CH_3 or C_2H_5. The alkenes themselves are also rapidly attacked by H, O and OH; in general, H atoms add to give the corresponding alkyl radical, while attack by O and OH yields acetyl radicals or acetaldehyde together with an alkyl radical. Thermal decomposition of the methyl and ethyl radicals formed by the initial attack on the fuel, and also in subsequent steps, competes with recombination and with oxidation by O atoms and molecular oxygen. This reaction scheme accurately reproduces the composition profiles through several lean and moderately rich alkane flames and also predicts the way flame speed varies with mixture composition [125, 126].

The reaction zone proper terminates in a type of quasi-equilibrium region with radical–radical reactions of the type

$$O + OH \rightarrow O_2 + H \qquad (6.69)$$

balancing the reverse branching reactions. However, true equilibrium has not yet been attained because the radical concentrations are still very high and the oxidation has proceeded mainly to carbon monoxide rather than carbon dioxide. The post-flame or recombination zone is a more extended region in which the slower recombination reactions occur, leading to further release of enthalpy, and carbon monoxide is oxidized to carbon dioxide if the fuel:oxygen ratio permits. The dominant reactions in the recombination zone of rich H_2–O_2 flames are

$$H + H + M \rightarrow H_2 + M \tag{6.70}$$

$$H + OH + M \rightarrow H_2O + M \tag{6.71}$$

and, in lean flames, they probably include

$$H + O_2 + H_2O \rightarrow HO_2 + H_2O \tag{6.72}$$

In hydrocarbon–oxygen flames, the final composition of the burnt gas is largely determined by the water-gas reaction

$$H_2 + CO_2 \rightleftharpoons H_2O + CO \tag{6.73}$$

The rapidity with which this reaches equilibrium depends primarily on the rate of the reaction

$$OH + CO \rightarrow CO_2 + H \tag{6.68}$$

Most of these features can be illustrated by discussing a system whose behaviour has already been dealt with under low-temperature conditions: the oxidation of methane.

6.5.4 The high-temperature oxidation of methane

In the self-propagating flame, transport of radicals ahead of the reaction zone means that an initiation step, such as

$$CH_4 + O_2 \rightarrow CH_3 + HO_2 \tag{6.37}$$

is not required and, furthermore, that HO_2 radicals are not necessarily involved in the reaction. As before, the major propagation reactions will involve radical attack on the fuel and

$$OH + CH_4 \rightarrow H_2O + CH_3 \tag{6.51}$$

$$H + CH_4 \rightarrow H_2 + CH_3 \tag{6.74}$$

$$O + CH_4 \rightarrow OH + CH_3 \tag{6.75}$$

will all be important.

The methyl radicals formed then react to give formaldehyde. Two mechanisms have been suggested, either reaction with molecular oxygen as before

$$CH_3 + O_2 \rightleftharpoons CH_3O_2 \quad (6.41)$$

$$CH_3O_2 \rightarrow CH_2OOH \quad (6.49)$$

$$CH_2OOH \rightarrow HCHO + OH \quad (6.50)$$

or reaction with oxygen atoms which are certainly present under flame conditions

$$CH_3 + O \rightarrow HCHO + H \quad (6.76)$$

The situation regarding the oxidation of methyl radicals is extremely complex and on the basis of the evidence obtained so far, it is difficult to decide which of these pathways is dominant.

The concentration of formaldehyde in the flame is extremely small, suggesting that its removal by radical attack is efficient.

$$H + HCHO \rightarrow H_2 + CHO \quad (6.77)$$

$$OH + HCHO \rightarrow H_2O + CHO \quad (6.54)$$

Formyl radicals then yield CO, either by decomposition

$$CHO\ (+M) \rightarrow CO + H\ (+M) \quad (6.57)$$

or by reaction with oxygen molecules or hydrogen atoms

$$CHO + O_2 \rightarrow CO + HO_2 \quad (6.56)$$

$$CHO + H \rightarrow H_2 + CO \quad (6.78)$$

It is important to note that formaldehyde need not now lead to branching: instead chain branching is believed to occur by the sequence found in the hydrogen–oxygen reaction

$$H + O_2 \rightarrow OH + O \quad (6.65)$$

Hydrogen atoms are produced by Reaction 6.76 and by reactions such as

$$OH + CO \rightarrow CO_2 + H \quad (6.68)$$

As well as oxidizing, methyl radicals also combine to form ethane

$$CH_3 + CH_3 \rightarrow C_2H_6 \quad (6.79)$$

and in fuel-rich flames, this is the principal way in which methyl

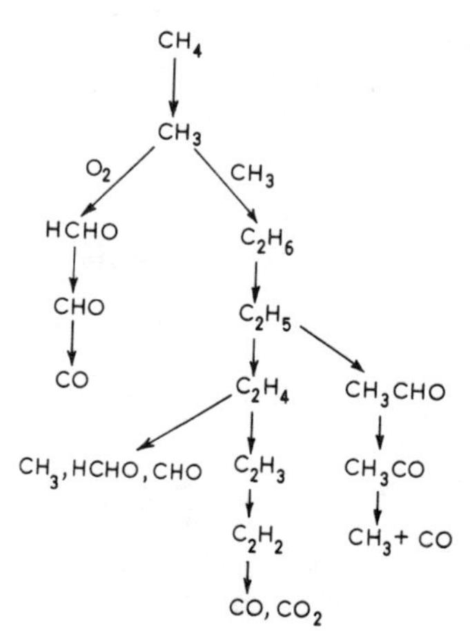

Fig. 6.8 Outline reaction scheme for the methane flame.

radicals disappear [125, 126]. Reactions of ethane lead to ethyl radicals, ethene and acetylene (ethyne). The complete reaction scheme is summarized in Fig. 6.8

It will be seen that the high-temperature reactions depend very much on the extent to which H atoms are present in the system and this in turn depends to a considerable extent on the fuel:oxygen ratio of the reactants. In the oxygen-rich case, the alkyl radicals will oxidize rapidly to formaldehyde etc. In the fuel-rich situation, stripping by H atoms becomes important and the radicals formed have a greater opportunity to react among themselves to give small quantities of higher hydrocarbons.

6.6 SUGGESTIONS FOR FURTHER READING

Ashmore, P.G. (1963) *Catalysis and Inhibition of Chemical Reactions*, Butterworths, London.

Fenimore, C.P. (1964) *Chemistry in Pre-Mixed Flames*, Pergamon, Oxford.

Minkoff, G.J. and Tipper, C.F.H. (1962) *Chemistry of Combustion Reactions*, Butterworths, London.

Semenov, N.N. (1958) *Some Problems in Chemical Kinetics and Reactivity*, Vol. 2, Pergamon, Oxford.

Shtern, V.Ya. (1964) *Gas Phase Oxidation of Hydrocarbons*, Pergamon, Oxford.

6.7 PROBLEMS

1. On the basis of the outline mechanism:

$$t\text{-Bu} + O_2 \rightarrow i\text{-}C_4H_8 + HO_2 \quad \text{(a)}$$

$$t\text{-Bu} + O_2 \rightleftharpoons t\text{-BuO}_2 \quad \text{(b)}$$

$$t\text{-BuO}_2 + t\text{-BuH} \rightarrow t\text{-BuOOH} + t\text{-Bu} \quad \text{(c)}$$

Show that

$$\frac{d[t\text{-BuOOH}]}{d[i\text{-}C_4H_6]} = \frac{K_b k_c [t\text{-BuH}]}{k_a}$$

and use the following data

$$k_a = 3.5 \times 10^9 \exp(-32\,\text{kJ mol}^{-1}/RT)\,\text{dm}^3\,\text{mol}^{-1}\,\text{s}^{-1}$$

$$K_b = 4.3 \times 10^{-6} \exp(113\,\text{kJ mol}^{-1}/RT)\,\text{dm}^3\,\text{mol}^{-1}$$

and

$$k_c = 10^9 \exp(-60.3\,\text{kJ mol}^{-1}/RT)\,\text{dm}^3\,\text{mol}^{-1}\,\text{s}^{-1}$$

to calculate the ratio of hydroperoxide to alkene expected in the products when the pressure of *i*-butane is 100 torr.

Are your results consistent with the observation that when *i*-butane is oxidized at 428 K, the products contain *t*-butylhydroperoxide and 1% *i*-butene, whereas at 573 K about 85% *i*-butene is formed.

2. After an initiating step producing alkyl radicals R, the main linear chain leading to products in combustion reactions can be written

$$R + O_2 \rightarrow A + HO_2 \quad \text{(a)}$$

$$R + O_2 \rightleftharpoons RO_2 \quad \text{(b), (-b)}$$

$$RO_2 + RH \rightarrow ROOH + R \quad \text{(c)}$$

$$RO_2 \rightleftharpoons QOOH \quad \text{(d), (-d)}$$

$$QOOH \rightarrow B + X \quad \text{(e1)}$$

$$QOOH \rightarrow C + X \quad \text{(e2)}$$

$$QOOH + O_2 \rightarrow D + X \quad \text{(e3)}$$

$$X + RH \rightarrow R + XH \quad \text{(f)}$$

where A is the conjugate alkene, and B and C represent stable molecular products. X is a chain-carrying radical such as OH, in which case XH = H_2O. Write down the differential equations for the formation of the radical species RO_2 and QOOH. Then using the stationary-state hypothesis for radical concentration show that

$$\frac{d[B]}{dt} = \frac{k_b k_d k_{e1} [R][O_2]}{\{(k_{-b} + k_c[RH])(k_{-d} + \Sigma k_e) + k_d \Sigma k_e\}}$$

where $\Sigma k_e = k_{e1} + k_{e2} + k_{e3}[O_2]$.

Assuming that Reaction b is equilibrated (i.e. $k_{-b} \gg k_c[RH]$ or k_d) and that k_{-d} is negligible, show that this expression reduces to

$$\frac{d[B]}{dt} = k_d K_b \boldsymbol{F}[R][O_2]$$

where $\boldsymbol{F}$ is the fraction of QOOH forming B.

3. The initial products of neopentane combustion at 753 K include *i*-butene, acetone and 3, 3-dimethyloxetan (DMO). The suggested mechanism for the formation of these products is

$$C_5H_{11} \rightarrow i\text{-}C_4H_8 + CH_3 \quad \text{(a)}$$

$$C_5H_{11} + O_2 \rightleftharpoons C_5H_{11}OO \quad \text{(b), (}-\text{b)}$$

$$C_5H_{11}OO \rightarrow C_5H_{10}OOH \quad \text{(c)}$$

$$C_5H_{10}OOH \rightarrow DMO + H \quad \text{(d)}$$

$$C_5H_{10}OOH + O_2 \rightarrow OOC_5H_{10}OOH \quad \text{(e)}$$

$$OOC_5H_{10}OOH \rightarrow CH_3COCH_3 + 2HCHO + OH \quad \text{(f)}$$

$$OH + C_5H_{12} \rightarrow H_2O + C_5H_{11} \quad \text{(g)}$$

Show that on this basis, both $[CH_3COCH_3]/[DMO]$ and $[DMO] + [CH_3COCH_3]/[i\text{-}C_4H_8]$ should be proportional to the concentration of oxygen in the reaction mixture (assume $k_{-b} \gg k_c$).

4. (a) Compare the rates of the possible internal hydrogen abstractions in the two alkylperoxy radicals derived from *i*-butane

$$CH_3-\underset{\underset{CH_3}{|}}{\overset{\overset{CH_3}{|}}{C}}OO \quad \text{and} \quad CH_3-\underset{\underset{CH_3}{|}}{\overset{\overset{CH_2OO}{|}}{C}}H$$

at 300, 500 and 800° C. What are the likely molecular products formed from these radicals?

(b) Make the same comparisons and predictions for the three possible *n*-pentylperoxy radicals.

7

Special aspects of gaseous combustion

7.1 EMISSION OF LIGHT

7.1.1 Introduction

Almost all combustion phenomena are accompanied by light emission and although the majority of the information available has been gained from studies on flames, simply for convenience, similar effects are observed with most other systems.

If the flame were in complete thermodynamic equilibrium then the hot gases should emit the continuous radiation predicted by the Planck radiation law for the appropriate flame temperature. For a true black body, the emissivity would have a value of unity at all wavelengths. In flames, the emissivity is closer to zero in the ultraviolet and visible regions although broad infrared bands with emissivity values close to unity are to be found. Once solid particles appear in the flame, the emissivity becomes much closer to that of a black or 'grey' body with the emissivity showing only slight variation with wavelength. Continuum emission may also be observed from ion–electron, atom–atom or radical–radical recombination processes because the energy levels of the upper state involved in the transition are not quantized.

However, the majority of the gaseous species present in flames possess discrete energy levels. This fact, coupled with the absence of low-lying electronic energy levels in the stable products of hydrocarbon combustion, for example, H_2O, CO_2, CO and O_2 is responsible for the low intensity of radiation in the ultraviolet and visible regions. Some of the transient species do have more favourable levels, for example, OH, CH, CN, C_2, HCO, NH and NH_2, and discrete spectra of these intermediates can be detected.

7.1.2 Non-equilibrium energy distributions

The most interesting feature of emission spectra from flames is the degree to which non-equilibrium distributions of excited species are observed and it is worth while to begin by summarizing the possible causes.

Thermal equilibrium among electronically excited states will be maintained by binary collisions

$$A + M \rightleftharpoons A^* + M \tag{7.1}$$

The efficiency of energy transfer between translational and electronic levels is usually very low and the mechanism above normally involves a change in the internal energy of species M, vibrational energy transfer proving to be particularly efficient. The processes maintaining thermal equilibrium are second order whilst the loss of energy by radiation to the surroundings will be first order

$$A^* \rightarrow A + \boldsymbol{h}\nu \tag{7.2}$$

where $\boldsymbol{h}$ is the Planck constant and ν is the frequency.

Since radiative equilibrium with the surroundings will rarely be attained and energy is being continuously lost, a pressure must exist below which *radiative depopulation* of the upper states can occur. Calculations based on a typical radiative lifetime of *ca.* 10^{-8} s suggest that radiative deactivation will be unimportant for flames burning at atmospheric pressure, but depopulation is to be expected at pressures of about 10 torr. Some experimental evidence for this effect has been obtained [9, 127].

A more common source of non-thermal distributions arises from the occurrence of chemical excitation processes. In combustion systems, the high temperatures in the reaction zone lead to high concentrations of atoms or radicals and chemical equilibrium is attained only in the later stages where recombination reactions are dominant. Reactions of these active species can lead to the formation of products in excited states: this is known as *direct chemiluminescence*. Although the excited products may not be strong emitters they can exchange energy with other species which do radiate: in this case, the process is referred to as *indirect chemiluminescence*. A good example [128] of the latter is provided by diffusion flames of atomic sodium and chlorine in which vibrationally excited sodium chloride is formed

$$Cl + Na_2 \rightarrow NaCl^{\dagger} + Na \tag{7.3}$$

and this then transfers energy to sodium atoms

$$NaCl^{\dagger} + Na \rightarrow NaCl + Na^{*} \quad (7.4)$$

The detection of non-thermal radiation from a particular intermediate does not therefore necessarily imply that the species has been produced in a chemiluminescent reaction. The appearance of chemiluminescence requires that deactivation of the excited molecules is relatively slow. Thus the observation of a thermal distribution does not mean that chemiluminescent reactions are excluded and overpopulation of excited states may well be found if the pressure is reduced.

Non-thermal distributions are not limited to electronic level populations. It is convenient to distinguish between electronic, vibrational and rotational energies and to define a 'temperature' based on the distribution among the appropriate energy levels. The measurement of a vibrational or rotational temperature which does not coincide with the adiabatic flame temperature then also provides evidence for a chemiluminescent excitation mechanism.

A classification of chemiluminescent reactions may be made as follows [129]:

Two-body association $$A + B \rightarrow AB^{*} \quad (7.5)$$

Three-body association $$A + B + C \rightarrow AB^{*} + C \quad (7.6)$$

$$A + B + C \rightarrow AB + C^{*} \quad (7.7)$$

Exchange reactions $$AB + C \rightarrow A^{*} + BC \quad (7.8)$$

$$AB + CD \rightarrow AC^{*} + BD \quad (7.9)$$

Although termolecular reactions are of little importance in combustion systems, the mechanism of three-body association normally involves an initial two-body association giving a complex with a relatively long life (a 'sticky' collision). This two-step mechanism can then prove of comparable efficiency to direct two-body association. The exchange reactions are more probable on kinetic grounds but, since a net change in the number of chemical bonds is not involved, few such reactions exist which can liberate large amounts of energy suitable, for example for electronic excitation.

Although the formal behaviour of chemiluminescent reactions is understood, it is difficult to ascribe mechanisms with any certainty in real systems. This is partly because absolute spectral intensity measurements are difficult to make, but mainly because the potential energy curves of the excited electronic states are simply not known.

Non-equilibrium distributions certainly occur in the OH band system; for example, the rotational temperature of $OH^*(^2\Sigma^+)$ has been measured [130] as 9000 K in an acetylene–oxygen flame at a pressure of 1.5 torr although the adiabatic flame temperature is only 3320 K. Vibrational non-equilibrium is observed in the reaction zone of the hydrogen–oxygen flame [131], the $v'=2$ and $v'=3$ levels being overpopulated with respect to the ground level. The mechanism of the over-population remains a matter of controversy although, in hydrocarbon–oxygen flames, OH* is almost certainly formed by the reaction

$$CH + O_2 \rightarrow CO + OH^* \tag{7.10}$$

The carbon monoxide–oxygen flame is rather better understood, the so-called 'flame' bands being attributed to the association

$$O(^3P) + CO(^1\Sigma^+) + M \rightarrow CO_2^*(^3\Pi) + M \tag{7.11}$$

$$CO_2^*(^3\Pi) \rightarrow CO_2(^1\Sigma_g^+) + h\nu \tag{7.12}$$

Direct association to $CO_2(^1\Sigma_g^+)$ would be exothermic by 530 kJ mol^{-1} and would violate the spin conservation rule so that the mechanism above is certainly to be expected. A third body is necessary if the $^3\Pi$ state is to be produced but some direct two-body association

$$O + CO \rightarrow CO_2^* \rightarrow CO_2(^1\Sigma_g^+)\ h\nu \tag{7.13}$$

also occurs, producing a continuum down to 250 nm. An appreciable number of the CO_2^* molecules formed by association transfer energy to oxygen molecules

$$CO_2^*(^3\Pi) + O_2(^3\Sigma_g^-) \rightarrow CO_2(^1\Sigma_g^+) + O_2^*(^3\Sigma_u^-) \tag{7.14}$$

leading to emission from the Schumann–Runge system of oxygen. The detailed characteristics of the emission do differ somewhat depending on whether the stationary flame or the explosion is examined but the same basic processes appear to be involved.

Other common flame emitters, for example, CH, the Swan bands of C_2, HCO etc. may also be formed in chemiluminescent reactions and a number of possible sources have been suggested. There are at least two processes which lead to the formation of CH*. The main reaction is

$$C_2 + OH \rightarrow CO + CH^* \tag{7.15}$$

and there is another process not dependent on C_2.
Among the reactions suggested are

$$O + C_2H \rightarrow CH^* + CO \tag{7.16}$$

and

$$O_2 + C_2H \rightarrow CH^* + CO_2 \tag{7.17}$$

The source of excited C_2 is still not established with certainty, but one possibility is

$$CH_2 + C \rightarrow C_2^* + H_2 \tag{7.18}$$

The spectrum of fuel-lean flames often contains features due to HCO^* which may be produced by

$$CH + HO_2 \rightarrow HCO^* + OH \tag{7.19}$$

In general, the experimental evidence is inadequate to provide convincing support for any particular mechanism.

Although the discussion has been concerned primarily with non-equilibrium light emission, it is worth mentioning that spectroscopic observations have been used to provide information on flame properties, particularly in the burnt gas region. The experimental procedure usually involves the addition of known quantities of metallic salts to the reactants thereby enabling the gas temperature to be measured by line-reversal techniques and concentrations of particular intermediates to be determined by monitoring the position of particular equilibria spectroscopically.

7.2 IONIZATION

7.2.1 Equilibrium ionization

The majority of combustion phenomena involve some degree of ionization and the presence of ions in a flame may be demonstrated by the distortion which occurs when an electric field is applied across it.

In general, ionization is not readily detected and special techniques are required for its study. The most common approach has been to introduce a pair of electrodes into the gas and measure the current flowing when a potential difference is applied across them. The actual measurement of an ion concentration is more difficult because the mobilities of the ions must also be determined and because the current observed depends on space charge and boundary layer effects. Ionization has also been followed by measurements of microwave attenuation and by direct detection of positive ions (and occasionally negative ones too) by sampling the gases from a flame into a mass spectrometer.

The ionization produces mainly positive ions and free electrons, although some electron capture may occur to give negative ions. The basic cause is, of course, the high temperature involved, the equilibrium constant for the process

$$A \rightleftharpoons A^+ + e \tag{7.20}$$

where A is an atom, being given by the Saha equation:

$$K = \frac{n_+ n_e}{n} = \left(\frac{g_+ g_e}{g}\right)\left(\frac{2\pi m_e \boldsymbol{k} T}{\boldsymbol{h}^2}\right)^{3/2} \exp\left(\frac{-E_I}{\boldsymbol{k} T}\right) \tag{7.21}$$

where K is the equilibrium constant, n_+, n_e and n are the numbers of positive ions, electrons and neutral atoms respectively per unit volume, g_+, g_e and g are their electronic degeneracies or statistical weights, m_e is the mass of the electron, $\boldsymbol{k}$ is the Boltzmann constant, T is the absolute temperature, $\boldsymbol{h}$ is the Planck constant and E_I is the ionization potential for species A.

Because of electron spin, $g_e = 2$ while g_+ and g depend on the electronic state of the ground state of the atomic ion and the atom, respectively, so that $g = (2J + 1)$ where J is the quantum number describing the total electronic angular momentum of the species. Some values of g and g_+ are given in Table 7.1, together with the corresponding values of E_I.

The degree of ionization depends very critically on the value of the ionization potential and the total ion density is therefore very sensitive to the nature of trace quantities of any easily ionized impurities which may be present. All the common flame gases and most of the radicals, for example, O_2, N_2, H_2, H_2O, CO, CO_2, OH and H, have high ionization potentials (see Table 7.2) and normally make only a negligible contribution to the total ionization. For example, CO will provide only 10^6 ions cm^{-3} at 3000 K. The commonest species with a low ionization potential (9.26 eV) found in flames is nitric oxide and, in

Table 7.1 Values of E_I, g and g_+ for some common metals [9].

Element	E_I(eV)	g	g_+
Li	5.39	2	1
Na	5.14	2	1
K	4.34	2	1
Mg	7.64	1	2
Ca	6.11	1	2
Al	5.98	6	1

Table 7.2 Ionization potentials of some typical flame components [9].

Species	E_I(eV)	Species	E_I(eV)
O_2	12.1	OH	13.2
H_2	15.4	NO	9.3
N_2	15.6	NO_2	9.8
CO	14.0	C_2H_2	11.4
CO_2	13.8	CH_4	12.9
H_2O	12.6	CH_3	9.8
H	13.5	CHO	9.8

1% concentration, this will yield 10^{11} ions cm^{-3} at 3000 K. The picture is changed drastically in the presence of metallic impurities. One part in 10^8 of potassium will give more ions than the 1% concentration of nitric oxide. In any combustion system the presence of minute traces of metallic impurities, which have no significant effect on any other aspects, completely dominates in the formation of ions.

Even in the absence of impurities, ions do occur in hydrocarbon flames. The total ion concentration, around 10^{11} ions cm^{-3}, far exceeds that expected on equilibrium grounds and the origin of the ionization remained a puzzle for many years.

7.2.2 Mechanism of ion formation

Mass spectral analysis has shown that in hydrocarbon flames the predominant positive ion is usually H_3O^+ although a large range of polymeric ions $C_nH_n{}^+$ and small concentrations of HCO^+ are also observed together with several negatively charged species; free electrons are also present. The levels of ionization are greatest in the reaction zone and fall off in the burnt gases.

It is now accepted that chemi-ionization occurs close to the reaction zone as a result of

$$O + CH \rightarrow HCO^+ + e \tag{7.22}$$

although ions are also produced by

$$CH^* + C_2H_2 \rightarrow C_3H_3^+ + e \tag{7.23}$$

particularly in fuel-rich systems.

The most important process following Reaction 7.22 is

$$HCO^+ + H_2O \rightarrow H_3O^+ + CO \tag{7.24}$$

in which a proton is transferred to a molecule such as H_2O with a proton affinity greater than that of CO. This ion then exchanges protons with species of higher proton affinity resulting in the formation of a wide range of ions of different structures.

Reaction of $C_3H_3^+$ with hydrocarbon molecules leads to the large polymeric ions which are probably associated with incipient soot formation (see later).

In the reaction zone radical concentrations frequently exceed the equilibrium values and thus ion concentrations greater than those corresponding to equilibrium are also produced by Reactions 7.22 and 7.23. The 'concentration overshoot' is thereby converted to an 'ionization overshoot' which persists into the postflame region (Fig. 7.1) where the major recombination reaction is

$$H_3O^+ + e \rightarrow H + H + OH \tag{7.25}$$

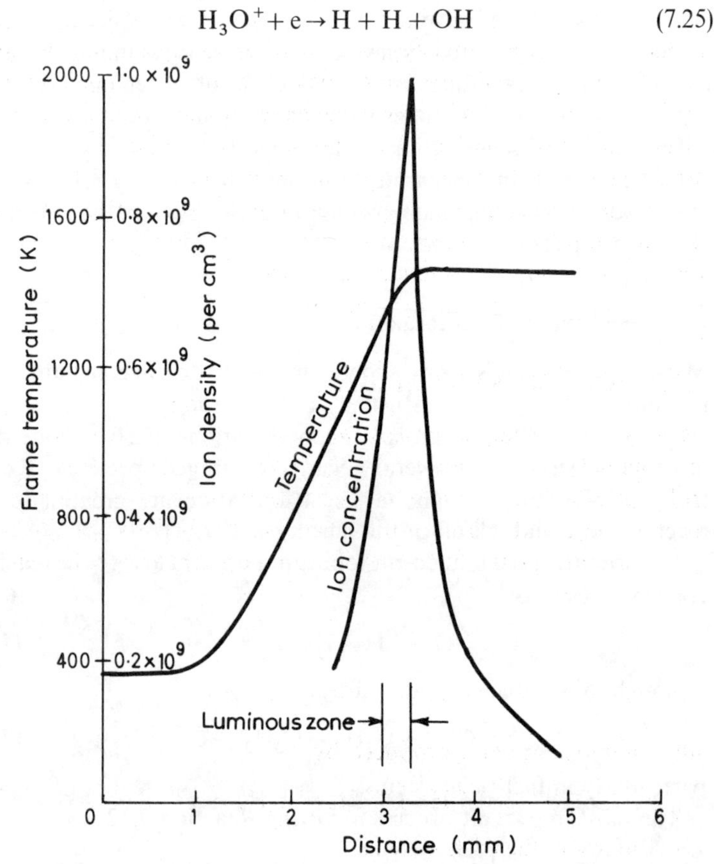

Fig. 7.1 Ion density profile through flame (after Calcote and King [132]).

with some contribution from

$$H_3O^+ + e \rightarrow H_2O + H \tag{7.26}$$

Negative ions are also observed in flames. The initial reaction is

$$O_2 + e + M \rightarrow O_2^- + M \tag{7.27}$$

followed this time by electron transfer to species of high electron affinity, for example, CH, O, OH, O_3, C_2H, CH_3O_2 and HO_2. Ions like O^- and OH^- are strong bases and abstract protons, for example,

$$CH_3OH + OH^- \rightarrow H_2O + CH_3O^- \tag{7.28}$$

Negative ions are lost by dissociative attachment, for example,

$$OH^- + H \rightarrow H_2O + e \tag{7.29}$$

Ionization in flames is important in schemes for the direct generation of electricity from a flame by burning it between the poles of a magnet and extracting a current using electrodes at right angles to both the direction of motion of the flame and the magnetic field (MHD or *magnetohydrodynamic* generation of electricity). The concentration of ions in the flame is enhanced by the addition of an easily ionized element such as potassium when the ionization persists longer than natural ionization because of the relative slowness of

$$K^+ + e + M \rightarrow K + M \tag{7.30}$$

which requires the presence of a third body, M. With potassium as additive, ions originate via

$$K + H_3O^+ \rightarrow K^+ + H_2O + H \tag{7.31}$$

$$K + M \rightarrow K^+ + e + M \tag{7.32}$$

Somewhat strangely, the addition of a halogen to a flame containing an alkali causes even more ions to form through the following fairly rapid steps:

$$K + Cl \rightleftharpoons K^+ + Cl^- \tag{7.33}$$

$$Cl^- + H \rightleftharpoons HCl + e \tag{7.34}$$

$$HCl + H \rightleftharpoons H_2 + Cl \tag{7.35}$$

The net effect of these three reactions is

$$K + 2H \rightleftharpoons K^+ + e + H_2 \tag{7.36}$$

and as the concentration of H atoms is greater than the equilibrium value, the local equilibrium concentration of alkali ions is also increased, the halogen acting, in effect, as a catalyst to this scheme.

7.2.3 Aerodynamic effects

The presence of ions in flames accounts for the aerodynamic effects produced by imposed electric fields. Although these phenomena were initially regarded as scientific curiosities, electrical techniques are now used for controlling combustion. A wide range of effects can be observed, the most general feature being that the flame bends towards a negatively charged body. Limits of stability and flame speeds are also affected by applied electric fields. At first sight it may appear strange that these effects occur since both positive and negative particles are present and the effects on each might be expected to cancel. It seems that the free electrons which constitute the majority of the negative species have a much greater mobility and are rapidly collected at the positive electrode leaving the flame gas with a net positive charge which then causes movement towards the negative electrode. This effect is termed the Chattock electric wind [133].

These comments have referred primarily to normal 'hot' flames although cool flames have also been shown to produce ionization. Ionization levels are also high in detonations but here it seems that equilibrium is established much more rapidly, presumably because of the higher temperatures and pressures involved.

7.3 CARBON FORMATION

7.3.1 Nature of soot

Although the oxidation of organic compounds leads eventually to carbon monoxide and carbon dioxide, fuel-rich flames tend also to produce solid carbon. This causes the intense yellow coloration characteristic of such flames and may lead to the formation of carbon particles as soot. The radiation from such flames approaches black-body intensity and can therefore be put to advantage in industrial burners because of the increased efficiency of heat transfer. However, the formation of solid particles which can deposit on cool surfaces must be kept to a minimum. Soot is also an undesirable product of combustion in engines where some is deposited as solid carbon in the combustion chamber and some is emitted as solid particles in the exhaust. For all these reasons, much effort has been devoted to understanding the mechanism of soot formation and to devising ways in which it may be minimized.

The nature of the soot varies both chemically and physically with position in the flame. When first observed, the 'young' soot has a particle diameter of about 5 nm which then increases, mainly by agglomeration in the later stages, to 200 nm or more. During this process the radical character of the soot decreases. The concentration of soot particles lies typically in the range 10^6–10^9 per cm^3. The hydrogen content of soot is quite high, about 1% by weight or 12% of the number of atoms present, and corresponds approximately to the empirical formula C_8H. Examination under the electron microscope reveals the deposited carbon appears to consist of a number of roughly spherical particles strung together rather like pearls on a necklace. The particles are made up of crystallites measuring about 1.3 nm by 2.1 nm. X-ray analysis shows that the carbon atoms are in a graphitic structure with the planes parallel to each other. In contrast to graphite itself, the planes are placed randomly one above the other with the hydrogen atoms distributed between them [134].

7.3.2 Mechanism of formation

As in most other aspects of flame chemistry, equilibrium considerations do not strictly apply and the presence of soot is dictated by the balance between the rates of reactions forming soot and those which cause its oxidation. Clearly, these depend very much on the fuel:oxygen ratio and the effects observed will be different in pre-mixed and diffusion flames. The luminosity of the flame is characteristic of carbon formation even though luminous flames do not necessarily deposit soot. Either luminosity or soot deposition may be used as a criterion of carbon formation although, of course, luminosity will occur earlier during the combustion.

Across any plane through a diffusion flame, the fuel:oxygen ratio varies from pure oxygen to pure fuel and one might expect that carbon formation would occur within the flame at any height. In fact, carbon formation first appears at the tip and is always restricted to the upper portion of the flame. The carbon formation is drastically reduced when the flame becomes turbulent. It therefore appears that the development of an appreciable layer of products between the fuel and oxygen is essential if carbon is to be formed. The height at which carbon is observed will also be dictated to some extent by the time required for certain processes to occur.

Furthermore, in diffusion flames, the residence times are longer and the temperatures in the soot-forming zone are lower than in pre-mixed flames. Pyrolysis of the fuel molecules to heavier, less-volatile hydrocarbons takes place and these condense and then carbonize. The most significant factors which determine the sooting tendency of a diffusion flame appear to be flame temperature and fuel structure. Alkenes containing four or five carbon atoms soot more readily than most aliphatic fuels, including acetylene, and this has led to the suggestion that intermediates such as butadiene are important in carbon formation, at least in diffusion-controlled situations.

In pre-mixed flames, the fuel: oxygen ratio is constant and one would only expect solid carbon to appear when the oxygen content is insufficient for complete conversion of carbon to carbon monoxide. In practice, carbon formation can be detected at oxygen:carbon ratios appreciably greater than unity. This can be interpreted in terms of the equilibrium

$$2CO \rightleftharpoons CO_2 + C_{solid} \qquad (7.37)$$

This equilibrium will be well over to the right at low temperatures and to the left above 1000 K. The numerical value of the equilibrium constant is not particularly critical because, at low temperatures, the rate of approach to equilibrium will be slow, whilst, at high temperatures, the position of equilibrium is away from carbon formation. However, its mere existence explains why the oxygen:carbon ratio is not the sole factor, carbon appearing at oxygen:carbon values greater than unity and furthermore with a dependence on temperature.

Various criteria have been suggested as measures of the tendency of a particular fuel to produce carbon and these can be used to rank different fuels. The order varies to some extent depending on the criterion adopted but even more so on the type of flame employed. An example is provided by non-stationary flames which give different results from flames supported on burners. Thus, propagating flames in cyanogen–oxygen mixtures deposit soot quite easily although it is difficult to obtain carbon from pre-mixed or diffusion flames of cyanogen. This suggests that the relative rates of carbon formation and carbon removal differ in the two systems. In general, polycyclic aromatic hydrocarbons give soot more readily than alicyclics and the latter more readily than aliphatic hydrocarbons. Unsaturation and branching in paraffin chains seem to give a greater tendency to sooting although here the distinction is less sharp. It is therefore dangerous to infer from tendencies to sooting much about the actual mechanisms involved.

Various theories have been proposed to explain the formation of carbon but so far the evidence is inadequate to permit a clear distinction being made between them. However, certain aspects of the mechanism are becoming established. To begin with, three phases of reaction are apparent. In the first phase, reactions occur which lead to the formation of carbon nuclei. Once these appear, a whole range of species can condense on their surface and eventually react there to increase the size of the spherical particles to between 10 and 50 nm. The surface is very active in catalysing these processes due to the presence of radical centres, or free valencies. These particles then grow by aggregation to form chains. Only a very few of the species which react in the later stages are involved in the initial nucleus formation and it is this initial stage which proves of greatest interest.

Two further complications arise. The detection of a particular intermediate of appreciable molecular weight in the flame need not signify any degree of involvement and, in fact, the appearance of a significant concentration is more indicative of a side product of the process. Furthermore, it now appears unlikely that the same mechanism necessarily applies to all flames, and there is considerable evidence to show that the mechanisms are quite different in flames of aliphatic and of aromatic hydrocarbons.

The majority of the earlier theories recognized that three basic steps are involved: break-up of the fuel molecules to lower molecular weight fragments, polymerization of these fragments, and dehydrogenation, and treated these processes as sequential. More recent work suggests that the mechanism involves the simultaneous occurrence of all three. For polymerization to occur, the fuel must almost certainly break down to give unsaturated species and in order to account for the relative rapidity with which carbon is formed radicals or ions must be involved in the reactions concerned.

Beyond this point, the situation is less certain. Intermediates involved in carbon formation could well include polyacetylenes, polyalkenes, polycyclic aromatic hydrocarbons and alicyclic hydrocarbons, and it is likely that all of these species may be involved to a greater or lesser extent. The critical intermediates have to meet a rather unique requirement of high stability at flame temperatures combined with a high reactivity towards polymerization.

It has been proposed that with aliphatic fuels, carbon formation takes place principally by an initial pyrolysis of the fuel to acetylene followed by pyrolysis to polyacetylenes and polyacetylene radicals which grow in length and cyclize [135]. One difficulty with this theory is that acetylene flames themselves do not readily form soot, and while polyacetylenes have been detected in flames, their growth is not fast enough to account for soot formation and rearrangement to the aromatic graphite-like structure of soot would be slow [136].

An alternative theory involves polymerization via aromatic compounds. Although benzene flames produce large quantities of soot, there is not much evidence of significant amounts of aromatics in the flames of aliphatic fuels, and even in benzene flames the polycyclic aromatics which are found cannot be formed by condensation of intact aromatic rings and much of the soot has been formed via non-aromatic hydrocarbon intermediates [137].

None of the theories yet proposed involving only neutral species is free from objection and it is worth noting that soot formation is strongly influenced by applied electric fields [133]. If, for example, an electric field is applied so as to increase the residence time of positive ions in the flame, larger particles are produced than when the ions are in the flame for a shorter time. The nature of the soot produced is also greatly influenced by the addition of electrons to the flame, and the addition of alkali metal ions to the flame gases also has a profound effect on soot growth. These observations, combined with the inadequacies of theories involving only neutral species, have led to the suggestion that ion–molecule reactions may be involved [138].

The presence of HCO^+ and $C_3H_3^+$ in hydrocarbon flames has already been mentioned (p. 169). In rich flames, the dominant ion is $C_3H_3^+$ and its concentration falls rapidly at the critical equivalence ratio for soot formation, its place being taken by larger ions. It appears that $C_3H_3^+$ reacts with other neutral species to form larger ions, for example,

$$C_3H_3^+ + C_2H_2 \rightarrow C_5H_3^+ + H_2 \tag{7.38}$$

$$C_3H_3^+ + C_4H_2 \rightarrow C_7H_5^+ \quad \text{etc.} \tag{7.39}$$

These larger ions can grow

$$C_5H_3^+ + C_2H_2 \rightarrow C_7H_5^+ \tag{7.40}$$

$$C_7H_5^+ + C_2H_2 \rightarrow C_9H_7^+ \quad \text{etc.} \tag{7.41}$$

Large gaseous ions are known to rearrange very rapidly to their most stable structure which is, in many cases, a polynuclear aromatic. Two very important ions in soot-forming flames are $C_{13}H_9^+$ and $C_{19}H_{11}^+$ and these have the structures

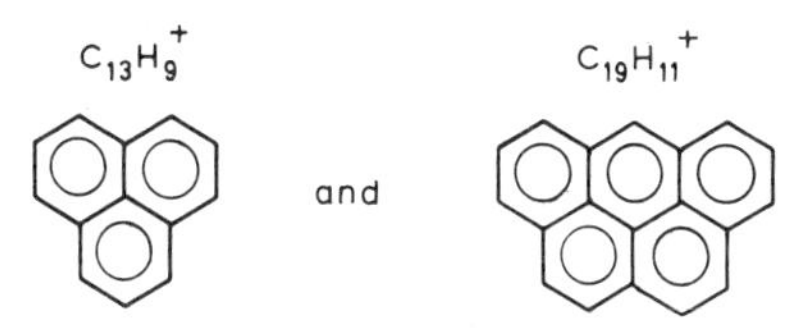

These ions will continue to grow and will eventually form soot particles. Although this theory cannot be regarded as established, it does explain a number of diverse observations whilst avoiding some of the difficulties of earlier theories.

7.4 COOL FLAMES, IGNITION AND MATHEMATICAL MODELLING

The appearance of cool flames and various types of ignition in the oxidation of organic compounds have already been mentioned in Chapter 6. These phenomena are sufficiently important and distinctive to merit further discussion.

Cool flames have been known for a very long time and they possess a number of characteristic features. While they are true combustion waves which can be stabilized in a suitable flow reactor, they travel at relatively low velocities (0.05–0.1 m s^{-1} compared with a normal flame speed of perhaps 0.4 m s^{-1}), the temperature rise is usually less than 150 K, the extent of reaction is low and the faint blue light which is emitted is characteristic of excited formaldehyde [139]. On average only one photon is emitted per 10^8 molecules of fuel consumed and the light emission is a side effect of the main chemical reaction. The origin of the emission is an electronically excited state of formaldehyde approximately 340 kJ mol^{-1} above the ground state. Only radical–radical reactions such as

$$CH_3O + OH \rightarrow HCHO^* + H_2O \tag{7.42}$$

and

$$CH_3O + CH_3O \rightarrow HCHO^* + CH_3OH \tag{7.43}$$

are sufficiently exothermic and both have been observed to give formaldehyde emission.

Most of the early work on cool flames was carried out in static systems (closed vessels) and in these both single and multiple cool flames may be observed. Sometimes one cool flame is quickly followed by ignition.

The first attempts at understanding cool flames ignored the temperature rise and reactant consumption. Although a later development incorporated thermal effects, none of these theories attempted to relate cool flames to other unusual features of hydrocarbon oxidation, nor did they invoke plausible reaction schemes. Temperature measurements in cool flame reactions in static systems established a connection between cool flames and the negative temperature coefficient of the rate and demonstrated that in the cool flame the temperature rise was sufficient to take the system into the regime where the rate fell as the temperature increased [140, 141]. Furthermore, mechanisms have been put forward which were qualitatively consistent with the observations [142]. These mechanisms incorporated the notion of a 'thermokinetic switch' which can best be explained by reference to an example. In acetaldehyde oxidation at about 250° C acetyl radicals are involved in the formation of the degenerate branching intermediate, peracetic acid,

$$CH_3CO + O_2 \rightarrow CH_3CO_3 \tag{7.44}$$

$$CH_3CO_3 + CH_3CHO \rightarrow CH_3CO_3H + CH_3CO \tag{7.45}$$

The branching reaction

$$CH_3CO_3H \rightarrow CH_3 + CO_2 + OH \tag{7.46}$$

causes the radical concentration to rise. Consequently, the rate of reaction increases and the rate of heat release is enhanced to the stage where a cool flame propagates. During the cool flame, the temperature rise results in the high activation energy reaction

$$CH_3CO + M \rightarrow CH_3 + CO + M \tag{7.47}$$

competing more favourably with Reaction 7.45 and this, together with the rapid increase in the rate constant for branching (Reaction 7.46), results in the concentration of peracetic acid diminishing to a very low value. The radical concentration and reaction rate then fall and the gas temperature drops back to that of the vessel. The peracetic acid can then build up again and, so long as there is an adequate supply of reactants, the sequence can be repeated. Ignition will occur if, during a cool flame, the temperature rises sufficiently for another reaction product (e.g. hydrogen peroxide or formaldehyde) to function as a branching intermediate.

The development of procedures for the numerical integration by computer of large sets of differential equations has encouraged attempts at mathematical modelling of combustion reactions with particular reference to modelling cool flames and ignition [143, 144]. While uncertainty regarding the detailed chemistry and lack of knowledge of many of the essential kinetic parameters has prevented complete success, progress has been made and a much better understanding of the basic mechanism has been gained. The combustion behaviour of hydrocarbons under conditions ranging from those at sub-atmospheric pressures in glass vessels to the high temperatures and pressures in internal combustion engines, can be simulated by a generalized mechanism based on a degenerate branching-chain scheme which contains the following essential features:

(a) The branching intermediate is formed in two ways, either by a radical + reactant process which runs parallel with the main propagation cycle, or by a radical + product process. The latter becomes more important in the later stages of reaction and is largely responsible for triggering the final hot ignition.
(b) Chain termination occurs by processes which are both first order and second order in radical concentration. It is the balance between these two which determines the shape of the cool-flame pulse.
(c) Heat is released in the chain-propagation steps.

This model displays all the commonly observed phenomena of slow combustion, single and multiple cool flames, negative temperature coefficient, and one-stage and two-stage ignition.

The starting point in modelling exercises of this type is the chemistry of the system. An alternative approach emphasizes the fact that only certain types of mathematical equations possess the distinctive features associated with cool flames, ignition and the like. In this way the unified theory outlined in Chapter 2 provides a framework for understanding complex combustion behaviour. Attention is focussed on the mathematical properties of the solutions of the simultaneous differential equations for the conservation of mass and energy, and the form of these solutions is related to various combustion phenomena [145, 146]. In doing this, the theory does not provide explicit information about the chemistry of the reaction but it does point the way to the design of better experiments which can be used to test the theory; these new experiments have also revealed novel types of behaviour.

The unified theory, like its predecessors [147, 148], is built on the assumption of spatially averaged concentrations and temperatures, and ignores reactant consumption. These assumptions are very crude when applied to closed vessels but may be much closer to the truth in a well-stirred flow system. For this reason much of the recent work on cool flames and ignition has been carried out in well-stirred flow reactors [149].

The skeleton reaction scheme written for hydrocarbon combustion is

$$A \rightarrow X \qquad \text{Initiation} \tag{7.48}$$

$$X + \cdots \rightarrow 2X \qquad \text{Branching} \tag{7.49}$$

$$X \rightarrow \text{inert} \qquad \text{Termination} \tag{7.50}$$

$$X \rightarrow \text{inert} \qquad \text{Termination} \tag{7.51}$$

where A is the fuel and X is the chain carrier. This branching-chain scheme does not explicitly consider any propagation processes, regarding them as thermoneutral. It involves a single chain carrier X which is destroyed by two separate processes, one having an activation energy less than that of the branching reaction, the other having an activation energy which is greater. These two alternative termination reactions are an essential feature of the mechanism.

The simultaneous differential equations governing the time variation of species concentration and temperature can be written down, the conditions under which oscillatory solutions exist are identified and the nature of the oscillations established for each solution. This is done by examining the behaviour of the system in the T–[X] (temperature–chain carrier concentration) phase plane. The thermokinetic steady states correspond to singularities in this phase plane (there are five for the particular formal scheme under discussion) and the nature of each singularity furnishes much information about the way the system approaches that particular steady state. In seeking the nature of the singularities, use is made of well-known results in stability theory [39, 40].

This formal scheme predicts that under certain conditions the system will approach a steady state via oscillations which may be damped or undamped. It also predicts one-stage and two-stage ignition and the existence of a negative temperature coefficient. With some modification, it can also explain multistage ignition. To do this, it is necessary to invoke the existence of a second intermediate, Y, which accumulates after successive cool flames, eventually reaching a critical value and then triggering the ignition.

In well-stirred flow reactors, the basic assumptions of the theory – uniform temperature and concentrations of major reactants – are closely realized and it is also possible to measure heat loss and heat generation terms quite readily. Experiments in well-stirred reactors have confirmed the predictions of the theory and a number of novel phenomena have been observed. In the acetaldehyde–oxygen system, nine chemically and physically distinct stable modes of reaction have been identified [149]. There are stable oscillations of seven clearly differentiated forms, and two stationary states. They can be grouped into five regimes:

1. Steady, slow reaction without light emission.
2. Stable oscillatory ignition with a simple waveform.
3. Stable oscillatory ignition with complex waveform (at least five different types were observed).
4. Stable oscillatory cool flames.
5. Steady rapid reaction accompanied by light emission (non-oscillating cool flames).

The temperature changes accompanying some of these phenomena are illustrated in Fig. 7.2.

Although the measurements of temperature change were supplemented by measurements of light emission and concentration of several important species, it has not yet been possible to link the experimental observations with the theory by means of a detailed chemical mechanism, nor has it been possible to identify with certainty the intermediates responsible for branching.

The oxidation of carbon monoxide also displays oscillations in both closed vessels and flow reactors. The study of this reaction is greatly complicated by the pronounced effect of water, hydrogen or hydrogenous impurities and by the difficulty in obtaining reproducible surface conditions.

In dry systems very long trains of oscillations can be generated in a properly conditioned vessel; often more than one hundred pulses of virtually constant amplitude can be observed. These oscillations are not multiple ignitions but isothermal light pulses and very little carbon monoxide is consumed at each excursion. The unified theory is capable of explaining the occurrence of this oscillatory behaviour either in terms of the 'dry' mechanism discussed earlier or using a more complex mechanism which includes

$$O + H_2O \rightarrow 2OH \tag{7.52}$$

followed by

$$CO + OH \rightarrow CO_2 + H \tag{7.53}$$

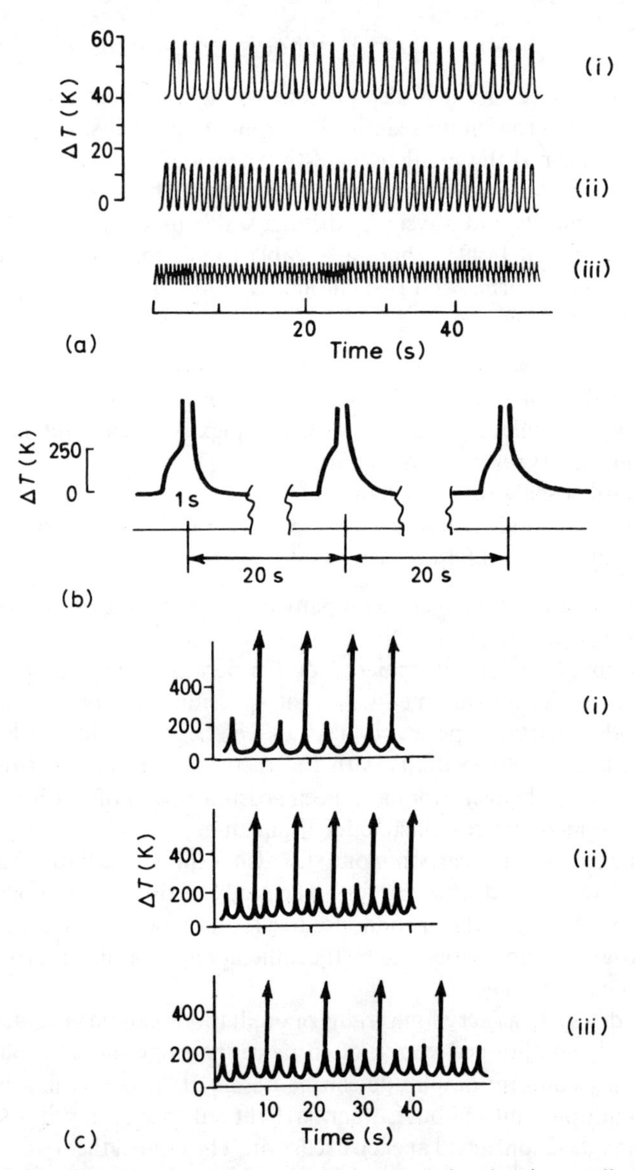

Fig. 7.2 Temperature changes in acetaldehyde combustion in a well-stirred flow reactor (after Gray *et al.* [149, 150]). (a) Sustained cool flame oscillations. Pressure, $16\,\mathrm{kN\,m^{-2}}$; residence time, 5s. The frequency increases as T_0 the vessel temperature is raised: (i) $T_0 = 588\,\mathrm{K}$, (ii) $T_0 = 593\,\mathrm{K}$, (iii) $T_0 = 600\,\mathrm{K}$. (b) Two-stage ignition. (c) Multi-stage ignition. Pressure, $20\,\mathrm{kN\,m^{-2}}$; residence time, 3s: (i) three-stage ignition, $T_0 = 500\,\mathrm{K}$ (two-stage ignition preceded by separate cool flame), (ii) four-stage ignition, $T_0 = 527\,\mathrm{K}$, (iii) five-stage ignition, $T_0 = 530\,\mathrm{K}$.

$$H + O_2 \rightarrow OH + O \tag{7.54}$$

$$H + O_2 + M \rightarrow HO_2 + M \tag{7.55}$$

and the removal of H, O and OH at the reactor walls [151].

7.5 SUGGESTIONS FOR FURTHER READING

Gaydon, A.G. and Wolfhard, H.G. (1979) *Flames, Their Structure, Radiation and Temperature*, 4th edn, Chapman and Hall, London.

Gaydon, A.G. (1974) *The Spectroscopy of Flames*, 2nd edn, Chapman and Hall, London.

Gray, B.F. (1975) Kinetics of oscillating reactions, in *Specialist Periodical Reports, Reaction Kinetics*, Vol. 1, The Chemical Society, London, p. 309.

Gray, P. and Sherrington, M.E. (1977) Self-heating, chemical kinetics and spontaneously unstable systems, in *Specialist Periodical Reports, Gas Kinetics and Energy Transfer*, Vol. 2, The Chemical Society, London, p. 331.

Kondratiev, V.N. (1964) *Chemical Kinetics of Gas Reactions*, Pergamon, London.

Laidler, K.J. (1955) *The Chemical Kinetics of Excited States*, Oxford University Press, Oxford.

Lawton, J. and Weinberg, F.J. (1969) *Electrical Aspects of Combustion*, Clarendon Press, Oxford.

Lewis, B. and Von Elbe, G. (1961) *Combustion, Flames and Explosions of Gases*, 2nd edn, Academic Press, New York.

Minkoff, G.J. and Tipper, C.F.H. (1962) *Chemistry of Combustion Reactions*, Butterworths, London.

7.6 PROBLEM

Assume that a typical flame gas at atmospheric pressure contains 15% H_2O, 0.5% H, 0.2% NO and 1 ppm (part per million) Na, and has a temperature of 2200 K. Use the Saha equation to calculate the equilibrium concentration of H_2O^+, H^+, NO^+ and Na^+ ions (neglect the statistical weight factor). Why is H_3O^+ a more common ion than H_2O^+ or H^+?

8 Combustion in mixed and condensed phases

In this chapter, we turn our attention to combustion phenomena in non-gaseous media. As regards pure phases, the behaviour in liquids is fairly similar to that in solids and, since very little practical use is made of combustion in liquids, the discussion is restricted to solids. Similarly, in the case of mixed phases, reactions between liquids and solids are relatively unimportant and we shall be concerned only with solid–gas and liquid–gas systems. It is assumed in discussing mixed-phase combustion that the condensed material plays the role of the fuel and the gas provides the oxidant.

8.1 REACTION IN SOLIDS

8.1.1 Introduction

Combustion reactions in solids lead to the same basic phenomena as are observed in gases, that is, explosions, in which the reaction may be considered essentially uniform over the whole volume, and propagating combustion waves, either detonations or deflagrations, which travel at a steady velocity through the material. As deflagrating combustion forms the basis of solid propellants used in rockets, and detonation is employed commercially in high explosives, the major technological aspects of these phenomena will be dealt with in subsequent chapters. Some of the more general features of reactions in solids will be introduced here.

In comparison with gases, solids possess thermal conductivities which are typically higher by an order of magnitude and diffusivities which are many powers of ten lower. The activation energy for diffusion is usually below $80\,kJ\,mol^{-1}$ so that, even at elevated temperatures, diffusion rarely becomes important until melting occurs. Mechanisms which depend on the motion of active centres through the solid lattice can usually be eliminated as they will be dominated in virtually all cases by mechanisms involving heat transfer. This means that the treatment of thermal explosions in Chapter 2 can be applied directly to reactions in solids.

From the Frank-Kamenetskii treatment (p. 32) it was possible to obtain the critical conditions for explosion applicable to various reaction cell geometries, and also relations for the pre-ignition temperature rise. These results can be applied directly to solids provided the material is confined in the same way. In practice, thermal explosions in solids are commonly generated when part of the surface is exposed to the atmosphere instead of being adjacent to a containing wall. This difference can be accommodated in the theory by the use of a suitable surface heat transfer coefficient. The simpler Semenov treatment of thermal explosions also happens to work quite well with solids; this is a consequence of the high thermal conductivity of solids which means that the major temperature gradients occur close to the boundaries rather than in the bulk material.

Although we are dealing with solid materials, the actual combustion rarely occurs in the solid phase. The temperature of the reaction zone is normally well above the melting point so that the solid can enter the zone as a liquid. Alternatively, the material may vaporize so that the combustion takes place in the gas phase or, in the case of a two-component mixture, at an interface between gas and solid. These considerations are virtually certain to apply in covalent solids but, in ionic materials, with high melting and sublimation temperatures, the situation is less clear-cut. In azides, for example, the mechanism is believed to involve loss of an electron from the N_3^- ion followed by reaction of two N_3 groups.

$$N_3^- \rightarrow N_3 + e \tag{8.1}$$

$$N_3 + N_3 \rightarrow 3N_2 \tag{8.2}$$

Since the heat is released in the later step and the two species have to diffuse together, reaction almost certainly occurs in the melt. In oxalates, on the other hand, the heat-evolving reaction is unimolecular and the ‘combustion’ can therefore occur directly in the solid phase

$$C_2O_4^{2-} \rightarrow C_2O_4 + 2e \tag{8.3}$$

$$C_2O_4 \rightarrow 2CO_2 \tag{8.4}$$

With ammonium perchlorate, which is an important constituent of propellants, the mechanism, at least in the absence of catalysts, involves proton transfer yielding adsorbed ammonia and perchloric acid which are then released into the gas phase. An exothermic reaction ensues and heat is conducted back to the ammonium perchlorate resulting in a temperature rise, accelerated decomposition, and eventually deflagration [152].

$$NH_4^+ + ClO_4^- \rightarrow NH_3(ads) + HClO_4(ads) \qquad (8.5)$$

$$NH_3(ads) \rightarrow NH_3(g) \qquad (8.6)$$

$$HClO_4(ads) \rightarrow HClO_4(g) \qquad (8.7)$$

$$NH_3(g) + HClO_4(g) \rightarrow \text{Complex mixture of products including nitrogen oxides, HCl } Cl_2 \text{ and } O_2 \qquad (8.8)$$

As might be expected from the range of products, the gas-phase reaction is very complex and involves several free-radical species [153].

Catalysts may influence the decomposition of perchlorates in various ways. Some, like magnesium oxide, react with ammonium perchlorate giving the corresponding metallic perchlorate which then forms a molten phase, while others, for example, copper (II) oxide, do not form a melt and reaction is limited to the catalyst surface. In either case, the essential feature of the catalysis is the enhancement of the rate of the heterogeneous decomposition of perchloric acid [154].

The effect of pressure on solid combustion reactions is difficult to predict. Since these reactions lead to gas evolution, one might expect that increase of pressure would inhibit the reaction. Although this is sometimes the case, increase of pressure also has the effect of restricting the hot products to the vicinity of the reaction zone, thus increasing heat transfer to the reactants and thereby the rate of reaction.

8.1.2 Detonation in solids

The major features of the detonation process have already been discussed with respect to gases: in solids and liquids, the situation becomes more complex both in terms of the phenomena observed and the underlying theory.

The first difficulty arises because of the higher densities and hence the higher pressures involved. For example, a typical explosive density of

1–2 g cm^{-3} (1000–2000 kg m^{-3}) can lead to pressures as high as 100 000 atm (10^4 MN m^{-2}). Although the products will be gaseous, ideal gas laws can no longer be applied, mainly because the effective volume occupied by the molecules themselves now becomes significant. This can be taken into account by using an expression of the form

$$p(V - b) = nRT \tag{8.9}$$

In the simplest such expression, the Abel equation, the *co-volume*, b, is taken as constant. This is satisfactory only at low densities and it is usually necessary to treat b as an empirical function of volume and temperature. Although other equations of state may be employed they are all basically empirical. This makes the prediction of detonation properties very difficult and, in practice, detonation measurements are used to determine the adjustable parameters in an empirical equation of state.

Ideal detonation is said to occur when the theoretical maximum velocity is attained. In practice, the observed detonation velocity may depend on the grain size, the dimensions of the *charge*, the mode of initiation etc. The ideal detonation velocity is therefore usually taken as the limiting value obtained as the diameter of the charge is increased, provided that the length is sufficient for true steady-state conditions to be established. Velocities in excess of the ideal value can occur if the wave is *overdriven* by using very energetic initiation, but they will eventually decay to the ideal detonation velocity. Lower velocities, corresponding to *non-ideal detonation*, are produced in more confined charges even after a steady state has been achieved. *Transient* and *dual-velocity* propagation (also termed *high-order* and *low-order* detonation) are not classified as non-ideal detonations since they can be attributed to non-steady state or instability effects. The ideal detonation velocity should depend only on the composition and density of the explosive. In cylindrical charges, non-ideal detonation, with a velocity below the maximum value, occurs between a *critical diameter*, d_c, below which detonation cannot be sustained and a *minimum diameter*, d_m, at which detonation becomes ideal.

Departures from ideal behaviour can be attributed in the main to the finite time required for reaction to occur. This leads in turn to a finite reaction zone length. Reaction zone lengths in solid-state detonations may vary from 0.1 to 10 mm. All solid explosives are composed of grains which may exist either as separate particles or as crystallites in a polycrystalline mass. According to the *surface-erosion* theory proposed by Eyring *et al.* [155], heat conduction into a grain will be insufficiently rapid to cause complete chemical reaction within the time scale of a detonation, even though the surface is directly exposed to the high-temperature products. This means that each grain must burn layer by layer from the surface inwards. It can be shown that the rate of reaction remains constant along a radius so that the time required for the grain to burn is simply proportional to its diameter.

Although the Chapman–Jouguet (C–J) plane normally coincides with the end of the reaction zone, this is not necessarily the case for long reaction times and the C–J condition may be achieved before reaction has ceased. Since energy released behind the C–J plane cannot contribute to the detonation process, the whole of the enthalpy is no longer available and a non-ideal detonation results. One effect is that the detonation velocity tends to fall with increasing particle size because of the thicker reaction zone.

A more serious effect occurs in unconfined charges. Since the products of reaction will expand laterally from the charge as well as in a rearwards direction, rarefaction waves will 'eat' into the reacting material behind the detonation front and, if they coalesce inside the reaction zone, the pressure and temperature at the C–J plane will be reduced and the detonation velocity will fall below the ideal value.

It is now possible to visualize the structure of the detonation process in an unconfined charge of solid explosive. Immediately following initiation, the detonation front appears spherical in shape with a centre at the point of initiation. In the early stages, the radius of curvature increases as the detonation propagates into the charge but it is found that the radius of curvature settles down quite abruptly to a steady value even before complete steady-state detonation is attained. The actual radius of curvature varies from one-half the diameter of the charge at the critical diameter, d_c, to about four diameters for $d \gg d_m$. The shape of the front at this stage is independent of the manner in which detonation was achieved. According to the *detonation-head model* [156], the lateral expansion waves give the reaction zone the shape of a truncated cone but, as the reaction zone grows, it eventually assumes a completely conical shape and steady-state detonation is achieved. The various stages of growth of the detonation head are depicted in Fig. 8.1.

The actual model of the detonation front requires some modification in solid explosives. The Zeldovich–von Neumann–Döring model [81–83] predicts the existence of a pressure spike between the shock front and the C–J plane. In solid explosives, this spike is considerably diminished and may disappear altogether. This may be attributed to

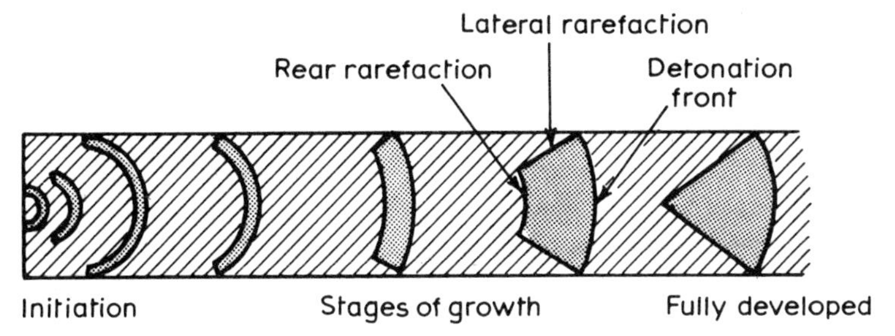

Fig. 8.1 Growth of detonation head in solid explosive (after Cook [156]).

the participation of heat transfer in the reaction zone. The 'heat pulse' is trapped in the zone so that the energy arises not simply from shock compression but also from heat conduction. The participation of heat transfer can be demonstrated by placing a layer of inert material in the explosive in such a way that the shock wave is allowed to pass whilst heat conduction is interrupted. Thin plates of material cause detonation to cease and require it to reform again on the opposite side.

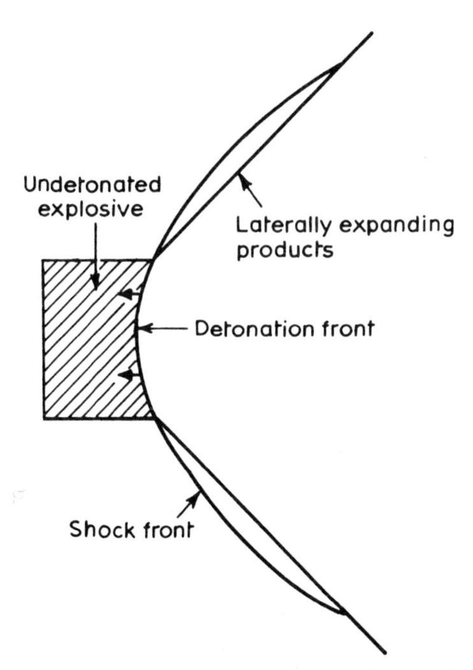

Fig. 8.2 Blast contour produced by a cylindrical charge of explosive away from the point of initiation.

Another rather interesting feature is the behaviour of the product gas temperature which has a maximum at the front and decreases through the reaction zone.

Behind the wave front of an unconfined charge, a *blast contour* extends laterally into the surrounding atmosphere (Fig. 8.2). Its leading edge is a shock wave and is usually marked by a bluish-white luminosity. Because of its lateral expansion, this shock wave decays rapidly and is overtaken by the expanding detonation products at about two charge diameters behind the front. The blast contour is therefore curved in the first region but becomes linear (conical) due to the constant lateral velocity of the products. For all explosives, the linear portion of the expanding wave makes an angle of about 36° to the axis of the cylinder. The colour in this region is normally brown. The expanding products eventually generate a further pressure or blast wave, distinct from the initial shock wave, which moves ahead of the products into the surrounding air.

Although we are not particularly concerned with detonation in liquids, the processes observed turn out to be qualitatively similar to those in solids. It appears that reaction occurs at the surface of gas bubbles in the liquid and propagates outwards at a constant rate. The role of the grain in solids is therefore taken over by the liquid between adjacent gas bubbles.

This discussion has provided an outline of the phenomena associated with the detonation of a cylindrical solid. The technology involved in the practical application of such charges is described in the following chapter.

8.1.3 Deflagration in solids

Even in solids, the detonation process depends mainly on the initial conditions and on the total enthalpy released by the reaction. In deflagrating combustion, on the other hand, the rate of burning depends on a complex interplay of different factors including the whole sequence of chemical reactions and a number of physical properties. For this reason, it is difficult to make many meaningful generalizations and, since deflagrating solids form the basis of solid propellants, a more detailed description of the effects observed with particular reactants has been deferred to Chapter 10.

One important distinction depends on whether the properties of both fuel and oxidant are contained in a single molecule or whether they are found in separate chemical species and hence as separate particles. In the latter case, proximity between fuel and oxidant is achieved by volatilization of one or both of the two reactants due to heat transfer from the reaction zone. The burning rate therefore depends on gas pressure, particle size etc., the rate-controlling process being primarily the rate of mixing.

When fuel and oxidant properties occur in the same species, mixing is no longer a problem and both reaction and vaporization can take place at the surface. The actual amount of reaction which occurs prior to vaporization rather than in the gas phase depends on the physical properties of the reactant, for example, its sublimation temperature. Even in the gas phase, separate reaction zones can commonly be distinguished.

Since solid explosives burn in layers parallel to a burning surface, the combustion is essentially a one-dimensional process. In double-base propellants (see p. 230), the following sequence of events may be distinguished [157] (Fig. 8.3). Immediately below the surface, in the *subsurface zone*, the reactants are heated by conduction, the rise in temperature causing decomposition of the explosive and also melting. At the surface, the material becomes volatile and burning continues in the so-called *'fizz'* reaction zone. About one-half of the available energy

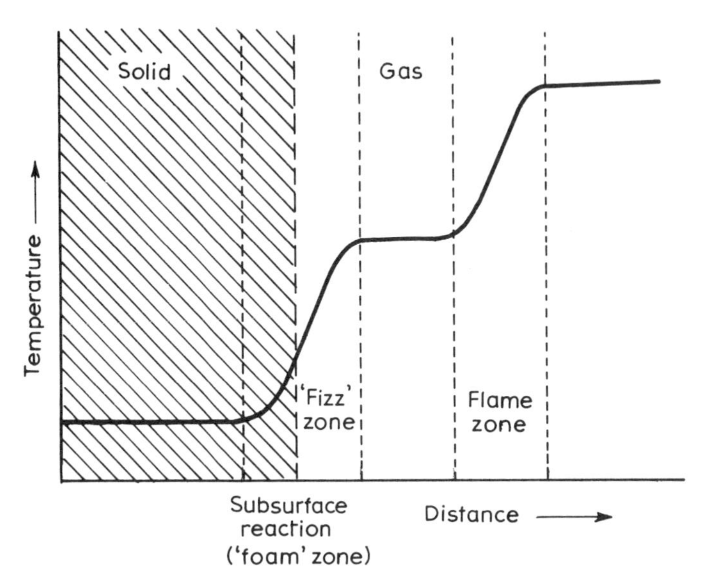

Fig. 8.3 Temperature profile and reaction zones in the burning of a solid propellant (after Huggett [158]).

is released in the fizz zone and after a short delay a luminous flame appears in which the reaction goes essentially to completion.

Various theories [158] have been proposed to correlate possible processes with the observed dependences on pressure, temperature and composition. The decomposition at the burning surface is largely a first-order process set in motion by heat conduction. The theories disagree to some extent on the relative amount of self-heating which occurs there, compared with simple volatilization of the fuel. The fizz zone is due to a thermal process in which reaction continues in the gas phase. The induction period between this region and the flame itself can be interpreted in two ways: either this is also a thermal reaction, but of higher order than the fizz zone, so that the two regions remain distinct, or, more probably, it corresponds to a branching-chain process in which critical concentrations of intermediates build up prior to reaction. The similarity between the behaviour here and the phenomena observed in two-stage ignition of hydrocarbon–oxygen mixtures may be noted.

Detailed and unequivocal mechanisms are not available for the whole burning process but a schematic representation is possible for the organic nitrates in the terms mentioned above.

Subsurface reaction

$$RCH_2ONO_2 \rightarrow RCH_2O + NO_2 \tag{8.10}$$

$$RCH_2O \rightarrow RCHO + RCH_2OH \tag{8.11}$$

Fizz reaction

$$RCHO + NO_2 \rightarrow HCHO + NO + H_2O + CO \tag{8.12}$$

Flame reaction

$$HCHO, NO, CO \rightarrow CO, CO_2, N_2, H_2, H_2O \tag{8.13}$$

8.2 REACTIONS IN MIXED PHASES

In solid–gas and liquid–gas systems, the behaviour observed is determined by three factors. The most fundamental is whether the actual combustion reaction occurs at the surface or in the gas phase. The heat released by the reaction will promote vaporization, melting and sublimation of the condensed material. In liquids, one may certainly claim that vaporization will always precede reaction so that combustion occurs in the gas phase. This is commonly found to be the case in solids as well. The second factor is the degree of dispersion, or the size of the particle in the gas. This factor determines the degree to which the vaporized fuel is able to mix with the gaseous oxidant. The final consideration is the relative motion of the two phases. In heterogeneous systems, the products of combustion tend to separate the fuel from the oxidant and this protection is reduced if the gas is in motion relative to the condensed material. Such motion will tend to increase the ease with which the fuel and oxidant come into contact but may, at the same time, reduce the rate of heat transfer to the fuel.

8.2.1 Burning of droplets

In dealing with gas-phase combustion, we shall mainly consider the example of a liquid fuel. The different situations which may occur are illustrated in Fig. 8.4 [159].

If the liquid is dispersed in very fine droplets, say below 10 μm in diameter, it will vaporize completely in the pre-heat zone and the resultant flame will be a typical pre-mixed flame, the reactants being intimately mixed within the reaction zone. The critical dimension below which a pre-mixed flame occurs depends on the rate at which the droplet is heated and this in turn depends on the final flame

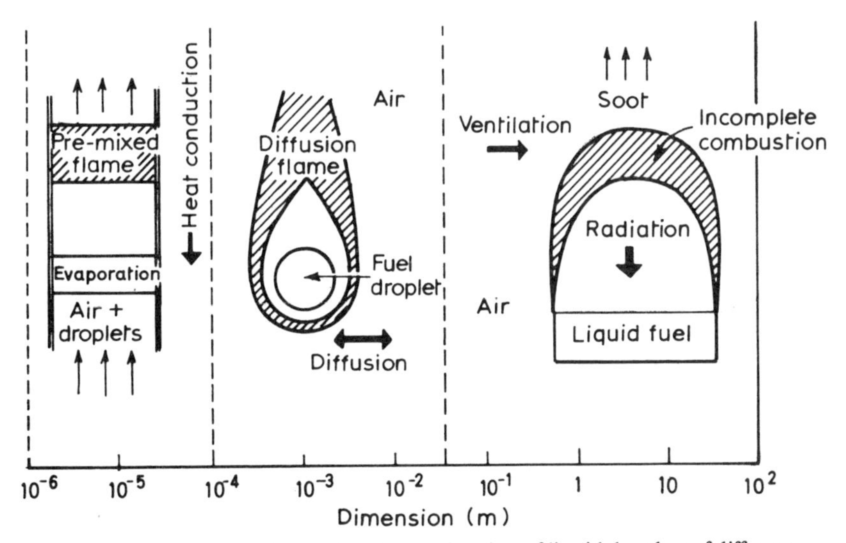

Fig. 8.4 Phenomena observed in the combustion of liquid droplets of different sizes. The heavy arrows indicate the major physical processes dominating the rate of burning (after Fristrom and Westenberg [159]).

temperature and the time which the particle spends in the pre-heat zone. The higher the burning velocity, the smaller the particles must be for complete vaporization and mixing. The critical size also depends on the latent heat of vaporization of the fuel (not the vapour pressure) and the thickness of the flame.

If the liquid droplet exceeds the critical size, but is less than about a millimetre in diameter, the combustion takes the form of a spherical diffusion flame around the droplet. At this stage the burning rate falls markedly and is determined by the rate of evaporation from the surface of the droplet. The behaviour is closely related to the diffusion flame formed on the wick of a paraffin lamp or a candle. The third factor, basically the degree of turbulence, obviously begins to play a part here since interdiffusion of reactants and convective heat transfer depend on such motion.

Most theoretical analyses of droplet burning assume that the liquid is at uniform temperature close to its boiling point. Fuel evaporates from the surface and diffuses into the surrounding atmosphere. A diffusion flame is established at the position where fuel and oxidant are in stoichiometric proportions and the heat released there is conducted back to the surface to maintain the temperature of the droplet and hence the supply of fuel by evaporation. Chemical reaction rate is assumed to be sufficiently fast for it to be unimportant and radiant heat transfer is neglected. The simple treatment which follows is that first given by Long [160].

In Fig. 8.5 a droplet, radius r_d, is surrounded by a spherical flame, radius r_f. The flame is assumed to have negligible thickness and the oxygen concentration at the flame surface is zero.

Consider unit mass of fuel reacting with n_{O_2} moles of oxygen to give n_{PR} moles of gaseous combustion products; the increase in the number of moles on combustion is $n' = n_{PR} - n_{O_2}$.

Since chemical reaction is assumed to take place instantaneously, at the surface of the droplet at any instant, the rate of consumption of oxygen is $4\pi r_d^2 \dot{m} n_{O_2}$, where $\dot{m}$ is the rate of evaporation of fuel in mass per unit area. This creates a net amount of material in the gas phase $4\pi r_d^2 \dot{m} n'$ moles which expands away from the droplet.

Thus, for a spherical shell, radius $r > r_f$, we can write a mass balance for oxygen

Diffusion in = Convection out + Consumption by reaction

$$4\pi r^2 \frac{\rho D}{M} \frac{dy}{dr} = 4\pi r_d^2 \dot{m} n' y \qquad + 4\pi r_d^2 \dot{m} n_{O_2} \tag{8.14}$$

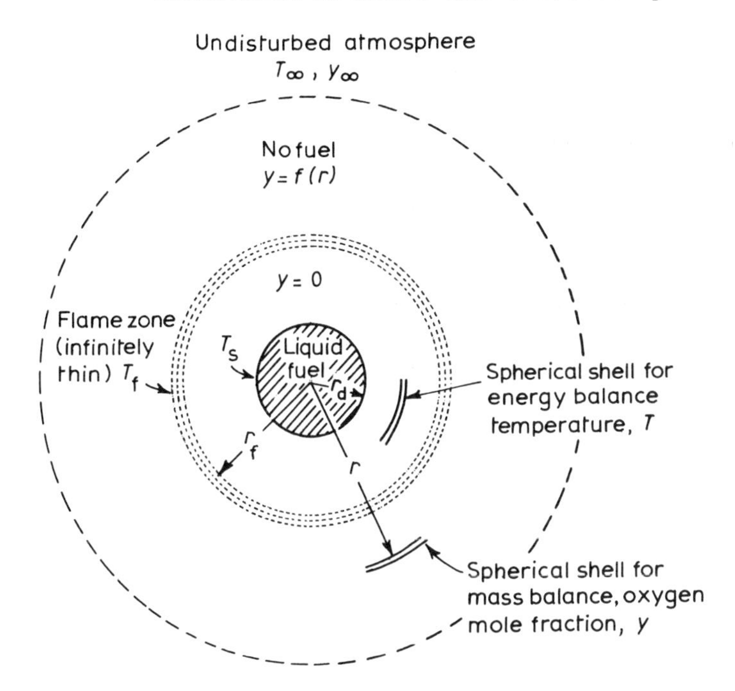

Fig. 8.5 The combustion of a liquid droplet.

where M is the molar mass of oxygen and ρ, D and y are, respectively, the density, diffusion coefficient and mole fraction of oxygen in the shell.

This can be rearranged to

$$\frac{\rho D}{r_{\mathrm{d}}^2 M} \frac{\mathrm{d}y}{(n'y + n_{\mathrm{O}_2})} = \dot{m} \frac{\mathrm{d}r}{r^2} \tag{8.15}$$

At a particular instant, $\dot{m}$ and r_{d} are fixed, while y varies with r. Assuming ρ and D can be assigned constant mean values then

$$\frac{\rho D}{r_{\mathrm{d}}^2 M} \int_{y_\infty}^{0} \frac{\mathrm{d}y}{(n'y + n_{\mathrm{O}_2})} = \dot{m} \int_{\infty}^{r_{\mathrm{f}}} \frac{\mathrm{d}r}{r^2} \tag{8.16}$$

and thus

$$\dot{m} = \frac{\rho D}{M n'} \frac{r_{\mathrm{f}}}{r_{\mathrm{d}}^2} \ln\left[1 + \left(\frac{n'}{n_{\mathrm{O}_2}}\right) y_\infty\right] \tag{8.17}$$

An energy balance allows r_{f}, which is difficult to measure, to be eliminated. Considering another spherical shell, this time within the flame so that $r_d < r < r_f$ and neglecting radiation, at a steady state,

Heat conducted in = Heat convected out

$$4\pi r^2 \lambda \frac{dT}{dr} = 4\pi r_d^2 \dot{m} l + 4\pi r_d^2 c_p (T - T_s) \quad (8.18)$$

where T is the temperature of the shell, T_s is the temperature of the surface of the droplet, λ is the mean thermal conductivity of the fuel vapour surrounding the droplet, c_p is the mean specific heat of the fuel vapour and l is the heat required to vaporize a unit mass of fuel at temperature T_s. Then

$$\frac{\lambda}{r_d^2} \int_{T_s}^{T_f} \frac{dT}{[l + c_p(T - T_s)]} = \dot{m} \int_{r_d}^{r_f} \frac{dr}{r^2} \quad (8.19)$$

whence

$$\dot{m} = \frac{\lambda}{r_d^2 c_p} \frac{\ln [1 + (c_p/l)(T_f - T_s)]}{(1/r_d) - (1/r_f)} \quad (8.20)$$

where T_f is the temperature of the flame and r_f its radius.

Eliminating r_f between Equations 8.17 and 8.20 it follows that

$$\dot{m} = \frac{\lambda}{r_d c_p} \left\{ \ln \left[1 + \left(\frac{c_p}{l} \right) (T_f - T_s) \right] + \frac{\rho D c_p}{M n' r_d} \ln \left[1 + \left(\frac{n'}{n_{O_2}} \right) y_\infty \right] \right\} \quad (8.21)$$

This expression comprises a heat transfer term and a mass transfer term. For hydrocarbon fuels, the former is much the more important and thus

$$\dot{m} \approx \left(\frac{\lambda}{r_d c_p} \right) \ln \left[1 + \left(\frac{c_p}{l} \right) (T_f - T_s) \right] \quad (8.22)$$

The rate of evaporation per unit area of a spherical drop, density ρ_d, is related to the decrease in radius by the expression

$$\dot{m} = -\rho_d \frac{dr_d}{dt} \quad (8.23)$$

Substituting for $\dot{m}$, and integrating, yields the result

$$d^2 = d_0^2 - \kappa t \quad (8.24)$$

Table 8.1 A comparison of experimental and calculated burning-rate coefficients [160].

Liquid fuel	κ (mm^2s^{-1}) Experimental value	Value calculated from Equation 8.25
i-octane	0.96	0.97
Cyclohexane	0.96	0.87
Benzene	0.97	0.97
Toluene	0.94	0.91
Xylene	0.88	0.89
Methanol	0.78	0.71
Ethanol	0.76	0.72

Note: In calculating κ, T_s has been taken as the boiling point of the liquid fuel and a mean value for λ of 0.126 $W\,m^{-1}K^{-1}$ has been assumed.

where d is the diameter of the droplet, d_0 being its value at $t = 0$ and

$$\kappa = \left(\frac{8\lambda}{\rho_d c_p}\right)\ln\left[1 + \left(\frac{c_p}{l}\right)(T_f - T_s)\right] \qquad (8.25)$$

Many experimental studies have shown that the diameter of burning droplets follows a relationship of the form of Equation 8.24 and κ is known as the *burning-rate coefficient.*

Table 8.1 shows that this simple theoretical treatment gives values for the burning-rate coefficient of a range of liquid fuels which are within 15% of those which have been determined experimentally even though the assumption of constant mean properties is obviously a crude approximation.

With the same basic assumptions of quasi-steady-state combustion, temperature-independent transport properties and an infinitely fast reaction (which implies an infinitely thin flame zone) more sophisticated treatments [59, 161] lead to the result

$$\kappa = \left(\frac{8\lambda}{c_p \rho_d}\right)\ln(1 + B) \qquad (8.26)$$

In this derivation, the Lewis number ($Le = \lambda/\rho c_p D$) has been taken as unity. The transfer number, B, is given by

$$B = \left(\frac{1}{l}\right)\left[c_p(T_\infty - T_s) + \frac{(-\Delta H)w_{OX,\infty}}{s}\right] \qquad (8.27)$$

where ΔH is the enthalpy of combustion, T_∞ and $w_{OX,\infty}$ are the temperature and the mass fraction of oxidant well away from the droplet, respectively, and s, the stoichiometric coefficient in terms of mass fractions, is w_{OX}/w_F. It also follows that

$$\frac{r_f}{r_d} = \frac{\ln(1+B)}{\ln[1+(w_{OX,\infty}/s)]} \tag{8.28}$$

In the case where evaporation takes place into hot combustion products, $w_{OX,\infty} = 0$ and thus

$$B = \left(\frac{1}{l}\right)[c_p(T_f - T_s)] \tag{8.29}$$

the two expressions for κ, Equations 8.25 and 8.26 then being equal.

Since the transfer number always enters into the expression for κ in a $\ln(1+B)$ term, κ is not very sensitive to changes in B and it is interesting to note that the burning rate coefficient can be calculated with reasonable precision using Equation 8.25 which is based on a simple model involving evaporation into a surrounding atmosphere at the flame temperature. It thus appears that for single droplets burning in free air, combustion is mainly controlled by heat transfer and the evaporation rate is only slightly in excess of that of a droplet evaporating into an infinite isothermal atmosphere at the flame temperature.

In practical combustion systems, droplets are usually burned in sprays and unless the spray is very dilute the single droplet model may not be applicable. Furthermore, in, for example, high-speed diesel engines spray combustion may take place under high pressures where the accuracy of the quasi-steady-state model is reduced. Theoretical treatments of droplet combustion under these more extreme conditions are available but further refinement is still needed before they can be regarded as completely satisfactory [162].

8.2.2 Pool fires

Beyond a droplet size of about a centimetre, the fuel behaves as a 'pool' and the phenomenon changes yet again. The situation here is that the rising volume of hot gas entrains air, restricts diffusion of reactants and reduces heat transfer by conduction (Fig. 8.6). The combustion zone becomes separated from the surface of the fuel and the temperature gradient is severely reduced so that eventually transport of heat occurs primarily by radiation. Because of the restricted access of the oxidant to the fuel the flame becomes very rich and, in the case of organic

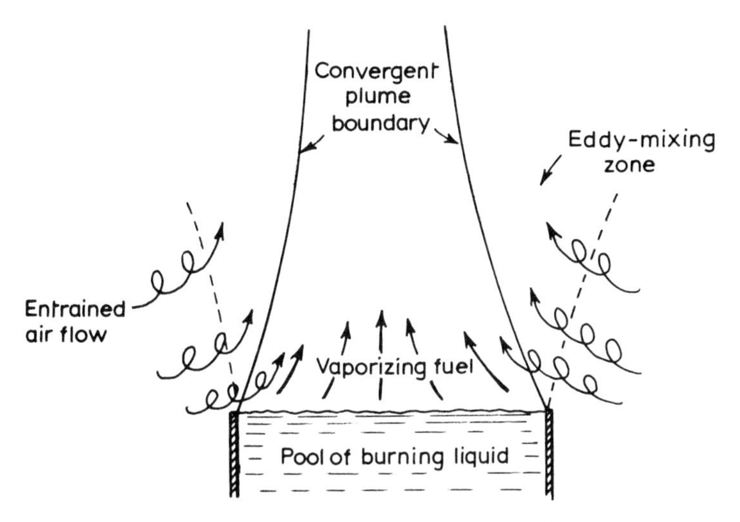

Fig. 8.6 A pool fire (after Herzberg [163]).

materials, soot is produced. The soot provides the radiation source which transfers heat to the fuel. Fires on large pools of fuel are very much affected by gas motion, that is, winds and, when the fire becomes large enough to produce its own meteorological effects, it is termed a *mass fire*. When motion in the surrounding air causes rapid rotation of the hot gases, a *fire whirl* may be set up.

For pool diameters of less than 50 mm the burning is usually laminar and the burning rate per unit area is roughly inversely proportional to the diameter of the pool. With larger pools, the burning rate is independent of diameter. Methanol fires produce relatively little radiation and with these the burning rate is constant with pools whose diameters are greater than 1 metre. For most liquid fuels, above diameters of about 0.3 m the radiative heat transfer and burning rate increase markedly with diameter.

In the intermediate region however, convective heat transfer is the controlling influence on the flame. It appears that low density flame gases rising into the denser ambient air above cause mixing of air with fuel vapour supplied by the pool surface. The resulting turbulent motion encourages convective heat transfer to the vaporizing surface and thus controls the rate of burning.

While a complete theoretical treatment of pool fires has yet to be achieved, some useful semi-empirical equations have been developed which make it possible to predict burning rates for pools of various liquids [164, 165].

The burning rate also depends on heat transfer within the liquid. The surface of a burning pool is usually just below its boiling point and it is sometimes possible to extinguish a fire by stirring liquid from below so that cool liquid of low vapour pressure is brought to the surface [166, 167].

The *fire point* is the lowest temperature of a liquid surface at which sustained diffusional burning of the liquid can occur and this is dependent on mass and heat transfer considerations which in turn influence the flame temperature [168].

8.2.3 Dispersions of solid particles

Solids which volatilize prior to combustion show very much the same dependence on particle size as liquids, but with an added effect of increased radiative heat transfer because of their high absorptivities. Dispersions of solid particles in air are occasionally used in industrial burners in the same way as liquid fuels. However, the combustion of solid dispersions is important because it constitutes a serious explosion hazard in factories and warehouses. Almost all solid materials which are endothermic with respect to their oxides, and this includes all organic materials and virtually all metals, can undergo a self-sustaining reaction or explosion when dispersed in fine particles. Dust explosions have actually been produced in several hundred different substances. The rate of reaction per unit mass of fuel and therefore the rate of heat release depends on the surface: volume ratio. Furthermore, the importance of radiative energy transfer means that virtually all these solid materials can give violent explosions at sufficiently small particle sizes. Consequently great care is required in handling powders, even such innocuous substances as flour, on an industrial scale. The situation is further complicated by the fact that in motion such materials rapidly build up a charge of static electricity and the discharge which may result can initiate the dust explosion.

In the case of metal particles, the oxidized products frequently have a higher sublimation temperature than the metal so that the products form a smoke which may be intensely luminous. This effect is exploited in photographic flash-lamps and in pyrotechnics. If the oxide products adhere to the surface of the solid, they can have the additional effect of protecting it against continued combustion.

Where the solid is present in bulk form, the geometry of the system tends to become critical. An isolated flat surface will not burn readily because of the energy lost by radiation. As soon as two surfaces lie opposite to each other the radiation becomes 'trapped' and combustion takes place far more readily. The amount of ventilation is also critical because of the need to replace the air consumed in the fire without at the same time increasing the loss of heat from the surface to the point where combustion is extinguished.

8.2.4 The combustion of polymers

The use of polymers in building materials and furniture has increased greatly in recent years and most of these materials can burn. When they do, the products of combustion are often toxic and there are many instances of fires in buildings where the burning polymers have added greatly to the damage caused and to the loss of life. The burning process is highly complex and chemical reactions may take place in three interdependent regions, namely, the condensed phase, the gas phase and the interface between them. In addition, some polymers, for example, polystyrene, pyrolyse with the formation of large amounts of volatile products which subsequently burn in the gas phase above the polymer; in this case, the presence of oxygen does not appear to have any effect on the breakdown of the polymer itself. With other polymers, for example, polypropylene, decomposition proceeds much more rapidly in the presence of oxygen and the gaseous products feed the flame formed by the oxygen-catalysed decomposition of the polymer.

Because of the safety implications there has been great interest in establishing criteria for the flammability of polymers and the most widely accepted measure is the *limiting oxygen index* (LOI) which is defined as the percentage of oxygen in the atmosphere just capable of supporting a flame. *Self-extinguishing polymers*, which do not continue to burn in air after being ignited thus have an LOI of greater than 21 and for such polymers to burn continuously the air must be enriched in oxygen.

In addition to burning or *flaming combustion*, some materials can undergo *smouldering combustion* which can become *glowing combustion*. In both cases, a front or wave involving gaseous oxidation of the polymer pyrolysis products propagates, through the solid. Glowing combustion is so-called because it is accompanied by the pale flame of carbon burning to carbon monoxide. A high proportion of fatalities in some fires has been due to the carbon monoxide produced by the smouldering combustion of certain polymers, especially polyurethane or neoprene foams.

The ease of polymer combustion is affected by many factors. The flammability is determined not only by the ease of thermal degradation of the polymer, but also by the nature and properties of the products. Transfer of the heat evolved in combustion back to the unburnt material is also important but the processes involved are so complex that to date no satisfactory correlations have emerged between polymer flammability and the parameters governing the separate stages of the burning process.

8.2.5 Heterogeneous combustion

In a large number of solid–gas systems, the actual combustion occurs on the surface, the oxidant being adsorbed there prior to reaction. The heat released then normally causes the products to be liberated rapidly as gases. Although the chemical reaction takes place on the surface, it is not necessarily the rate-determining step. The complete process takes place by a sequence of events and any one of these may, in principle, control the rate. The first step is the transfer of reactant (normally oxygen) from the bulk gas through the stagnant layer of gas adjacent to the surface of the solid particle. The reactant is then adsorbed and reacts with the solid before the gaseous products are released and diffuse away. If, as is frequently the case, the solid is highly porous then much of the available surface can only be reached by diffusion along the relatively narrow pores and this process may be rate controlling. The formation of a solid involatile combustion product (ash) may also complicate matters, but for most purposes we can disregard this. Control may thus be exercised by:

(a) Gas–film diffusion.
(b) Adsorption and chemical reaction, which are usually treated together under the heading of chemical reaction control.
(c) Pore diffusion.

Diffusion of products away from the surface is seldom rate controlling.

From this simple model it is possible to draw certain conclusions. Clearly, reaction control will occur if the surface reaction is slow compared with the diffusion processes. Since the surface reaction depends on the Arrhenius term, $\exp(-E/RT)$, while diffusion shows a less-marked temperature dependence, reaction control predominates at low temperatures while diffusion control becomes more important at higher temperatures.

Coal combustion provides a good example of the way the controlling process is influenced by the conditions under which reaction is taking place. After the initial stages of coal combustion during which volatile combustible material is given off, an involatile residue of porous carbon known as char, or coke, which comprises up to 90% of the original mass of the coal, is left behind; it is with the burning of this residual matter we shall now be concerned [169]. Three different regimes, or *zones*, of combustion can be distinguished [170].

In Zone I, the rate of diffusion to and away from the surface is very fast compared with the rate of the surface reaction. For a spherical particle the rate of reaction per unit external surface area is

$$\frac{-\mathrm{d}n_{O_2}}{\mathrm{d}t} = \frac{1}{6} S\rho d[O_2]^j A_s \exp(-E_s/RT) \tag{8.30}$$

where n_{O_2} is the number of moles of oxygen per unit area of surface, S is the external surface area per unit mass of carbon, ρ is the density of the particle, d is the particle diameter, j is the order of reaction, $[O_2]$ is the concentration of oxygen in the bulk gas, and A_s and E_s are the Arrhenius parameters of the surface reaction.

Zone I kinetics are observed at low temperatures. At very much higher temperatures, the rate at which oxygen molecules are transported from the bulk gas to the external surface is slow enough to be rate controlling and this is the region known as Zone III. Here the observed rate can be equated to the molar flux of oxygen to unit area of external surface. If the oxygen molecules react as soon as they reach the surface the rate (again per unit surface area) is given by

$$\frac{-\mathrm{d}n_{O_2}}{\mathrm{d}t} = \frac{k_g[O_2]}{\mu} \ln(1 - \mu y) \tag{8.31}$$

where y is the mole fraction of oxygen in the bulk gas and μ is a numerical factor arising principally because of the change in the number of gaseous moles in reaction; for $C + O_2 \rightarrow 2CO$, $\mu = -1$. The mass transfer coefficient, k_g, is given by

$$k_g = Sh\, D/d$$

where Sh is the dimensionless Sherwood number, D is the molecular diffusion coefficient of oxygen in the gaseous medium and d is, as before, the particle diameter.

The Sherwood number depends on the square root of both the mass flow rate and the particle diameter, and so the reaction rate increases with increasing mass flow around the particle but falls as the particle diameter increases. The effects of temperature on Sh and D combine to

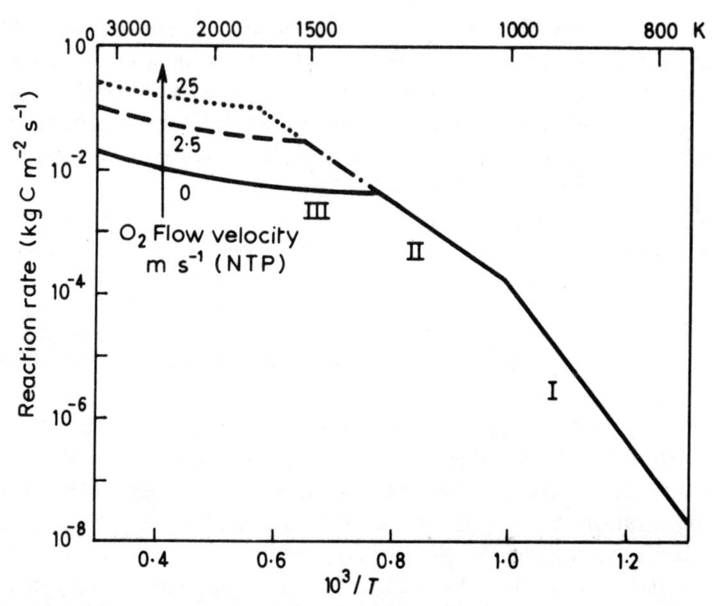

Fig. 8.7 The three zones in the reaction of a 10 mm sphere of pure, porous graphite with oxygen at 1 atm (after Mulcahy [170]).

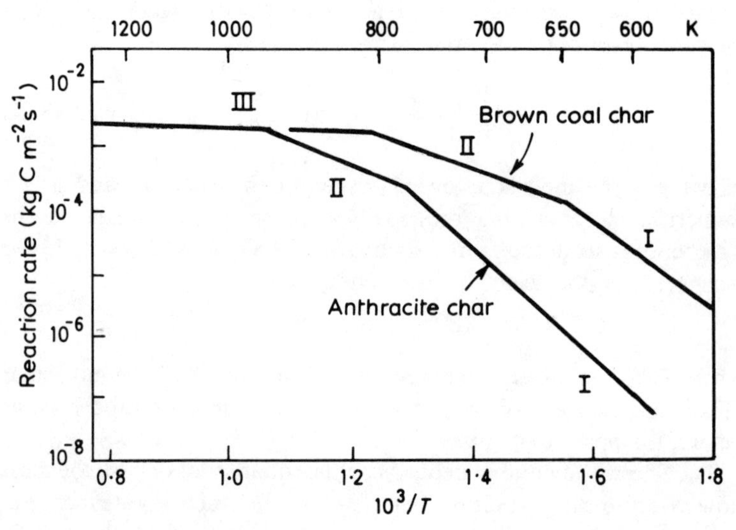

Fig. 8.8 The three zones in the reaction of 10 mm carbon spheres in still air (after Mulcahy [170]).

give a small increase in rate as the temperature rises. All these characteristics are markedly different from those in Zone I.

Zone II is intermediate between Zones I and III. Here, oxygen is transported quickly to the external surface, but diffuses relatively slowly down the pores before reacting. By considering a simplified model of the pore structure it can be shown [169] that the reaction rate per unit area of external surface is independent of the particle size while the apparent activation energy is half the activation energy of the true surface reaction (E_s) and the apparent order of reaction is greater than the true order (j).

Actual coal chars behave in precisely the way described and Fig. 8.7 illustrates the three regimes in the reaction of a porous graphite.

In practice there are enormous differences between the reactivities of different carbons. Some of these can be ascribed to variations in pore structure and some to the presence of impurities (e.g. alkali metal salts) which have a pronounced catalytic effect on the surface reaction. Consequently, the temperature ranges corresponding to the three zones differ (Fig. 8.8); while Zone I behaviour may be observed for a reactive brown coal char up to about 650 K, a less reactive anthracite char may follow Zone I kinetics up to nearly 800 K. At very high temperatures, both materials react at the same rate which is controlled by transfer of oxygen to their external surface (Zone III kinetics).

8.3 SUGGESTIONS FOR FURTHER READING

Bowden, F.P. and Yoffe, A.D. (1958) *Fast Reactions in Solids*, Butterworths, London.

Cook, M.A. (1958) *The Science of High Explosives*, Reinhold, New York.

Cullis, C.F. and Hirschler, M.M. (1981) *The Combustion of Organic Polymers*, Oxford University Press, Oxford.

Fristrom, R.M. and Westenberg, A.A. (1965) *Flame Structure*, McGraw-Hill, New York.

Gerstein, M. and Coffin, K.P. (1956) Combustion of solid fuels, in *Combustion Processes* (eds B. Lewis, R.N. Pease and H.S. Taylor), Oxford University Press, Oxford, p. 444.

Gray, P. and Lee, P.R. (1967) Thermal explosion theory, in *Oxidation and Combustion Reviews* (ed. C.F.H. Tipper) Vol. 2, Elsevier, Amsterdam.

Longwell, J.P. (1956) Combustion of liquid fuels, in *Combustion Processes* (eds B. Lewis, R.N. Pease and H.S. Taylor), Oxford University Press, Oxford, p. 407.

Palmer, K.N. (1973) *Dust Explosions and Fire*, Chapman and Hall, London.

Spalding, D.B. (1979) *Combustion and Mass Transfer*, Pergamon, Oxford.

Ubbelohde, A.R. (1956) Detonation processes in gases, liquids and solids, in

Combustion Processes (eds B. Lewis, R.N. Pease and H.S. Taylor), Oxford University Press, Oxford, p. 577.

Vulis, L. (1961) *Thermal Regimes of Combustion* (translated by M.D. Friedman), McGraw-Hill, New York.

Walker, P.L., Ruskino, F. and Austin, L.G. (1959). Gas reactions of carbon. *Advances in Catalysis*, **11**, 133.

8.4 PROBLEMS

1. Use Long's model to calculate the burning-rate coefficient of *n*-heptane and hence find the time for a droplet of initial diameter 1 mm burning in air (a) to decrease to one half its initial diameter, and (b) to disappear completely.

   Data:

   $\lambda = 0.115\,\mathrm{W\,m^{-1}K^{-1}}$ $\qquad c_p = 4600\,\mathrm{J\,kg^{-1}K^{-1}}$

   $\rho_d = 611\,\mathrm{kg\,m^{-3}}$ $\qquad l = 317\,\mathrm{kJ\,kg^{-1}}$

   $T_s = 372\,\mathrm{K}$ $\qquad T_f = 2305\,\mathrm{K}$

2. The latent heat of vaporization of benzene is $393\,\mathrm{kJ\,kg^{-1}}$, its boiling point is 80° C and the mean specific heat of its gaseous combustion products is $1.24\,\mathrm{kJ\,kg^{-1}K^{-1}}$. The enthalpy change in combustion is $\Delta H = -40.6\,\mathrm{MJ\,kg^{-1}}$. Calculate the transfer number B for benzene burning in (a) air, and (b) oxygen.
3. Use the value of B obtained in question 2 to evaluate the burning time and flame radius of a 50 μm diameter droplet of benzene burning in air, given that the density of liquid benzene is $880\,\mathrm{kg\,m^{-3}}$ and the thermal conductivity of the gases around the droplet is $0.0754\,\mathrm{W\,m^{-1}K^{-1}}$.

9

High explosives

9.1 INTRODUCTION

A general description of the physico-chemical phenomena accompanying combustion reactions in solids has been given in Chapter 8. In the present chapter we shall discuss some of the more technological aspects involved in the use of explosive materials.

Explosives can be subdivided into two main type: those which detonate, and are normally termed *high explosives*, and those which deflagrate. The latter are classified as *low explosives* in the United States and as *deflagrating explosives* in the United Kingdom. The peak pressure attained with a high explosive may be as high as 300 000 atm whilst that of a deflagrating explosive does not normally exceed 4000 atm. Comparable destructive power can of course be obtained from the latter by suitable confinement. Of the high explosives, some are detonated by all normal ignition sources, including heat, electric spark, flame, or mechanical impact, and are termed *primary explosives*. *Secondary explosives*, on the other hand, will only detonate under the influence of an externally applied shock or detonation wave and merely burn without detonating when ignited by a flame. Small quantities of primary explosives are used in devices called *detonators* to induce explosive reaction in bulk quantities of secondary explosives. The distinction is largely one of degree: the majority of explosives initially undergo a deflagration reaction which eventually accelerates to give a detonation. In the secondary explosive, a stronger energy source is required for ignition and the build-up to detonation takes much longer. Even deflagrating explosives may be made to detonate under appropriate conditions. Although the rate of energy release in deflagrating explosives is necessarily lower, the total amount of energy available is likely to be comparable with that released by high explosives. Both types of material may be considered for commercial operations such as blasting but the lower rate of burning in deflagrating explosives is an essential requirement for propellants. The energy available from, or work done by, a standard quantity of explosive is termed the *explosive strength* or, less desirably, *explosive power*. The explosive strength may be referred to unit weight (*weight–strength*) or unit volume (*bulk–strength*) of explosive.

9.2 EXPLOSIVE MATERIALS

Although the demands of any particular application may vary from a high-pressure explosive, usually for military purposes, to a low-pressure material to act as a propellant or a fuse, the number of explosive substances in common use is relatively small. The required properties are obtained from a suitable combination of explosives and non-explosive ingredients. A list of the commonest materials is given in Table 9.1.

The earliest explosive, 'blackpowder' or 'gunpowder', comprising potassium or sodium nitrate, sulphur and charcoal, is a deflagrating explosive and is used nowadays mainly in pyrotechnics (fireworks) and fuses. The first of the modern explosives, and possibly the most versatile, *nitroglycerine*, is a detonating material. Because pure nitroglycerine is a liquid and is extremely sensitive, that is, prone to accidental initiation, it is normally mixed with an absorbent to give a stable, dry explosive. The *straight dynamites* contain nitroglycerine, sodium or potassium nitrate, and a combustible material such as wood pulp or flour, normally in a combination such that the combustible material is fully oxidized by the nitrate. Another series of dynamites incorporates collodion cotton which dissolves in nitroglycerine to give a stiff, plastic substance very suitable for loading into a borehole. The three commonest blasting explosives – *blasting gelatin*, *gelatin dynamite* and *gelignite* – are all based on combinations of nitroglycerine with collodion cotton, woodmeal and potassium nitrate.

Another important explosive, *ammonium nitrate*, is used in mixtures with combustibles or with other explosives such as nitroglycerine. These mixtures are important because of their low explosion temperatures, which reduce their tendency to ignite the coal dust–air or methane–air mixtures that are found in coal mines (see later). Together with this feature, ammonium nitrate mixtures also have low sensitivity and are very safe to handle. It is at first sight surprising, therefore, to find that the most serious accidental explosions, resulting in the greatest loss of life, have been caused by ammonium nitrate.

Table 9.1 Properties of explosive compounds. (The densities quoted are normally those at which the detonation velocities have been measured.)

Name	Formula	Density ($g\,cm^{-3}$)	Detonation velocity ($m\,s^{-1}$)	Characteristics	Uses
TNT (trinitrotoluene)	CH_3, O_2N, NO_2, NO_2 (benzene ring)	1.63	6950	Safe to handle, Contains insufficient oxygen for complete combustion.	Popular as military and commercial explosive. Used in conjunction with ammonium nitrate– AMATOLS.
PETN (pentaerythritol tetranitrate)	O_2NOH_2C, CH_2ONO_2, C, O_2NOH_2C, CH_2ONO_2	1.77	8300	Very powerful but too sensitive. Expensive to manufacture.	Used as military explosive in combination with other explosives, e.g. with TNT– PENTOLITES.
RDX (cyclotrimethylene-trinitramine)	NO_2, N, H_2C, CH_2, O_2N-N, $N-NO_2$, CH_2	1.73	8500	Thermally stable and powerful. Requires some densitization.	Powerful military explosive. May be used with TNT.
HMX (tetramethylene-tetranitramine)	CH_2, O_2N-N, $N-NO_2$, CH_2, CH_2, O_2N-N, $N-NO_2$, CH_2	1.84	9124	High power, high stability.	Powerful military explosive.

Table 9.1 (*Contd.*)

Name	Formula	Density ($g\,cm^{-3}$)	Detonation velocity ($m\,s^{-1}$)	Characteristics	Uses
TETRYL (trinitrophenyl-methylnitramine)	$(NO_2)_3C_6H_2N(CH_3)NO_2$ (2,4,6-trinitrophenyl ring with $N(CH_3)NO_2$)	1.6	7500	Powerful. Sensitive to friction or percussion.	Military explosive. Used to prime less easily initiated explosive.
Ammonium nitrate, AN	NH_4NO_3	0.7–1.0	1100–2700	Not normally considered as an explosive when pure.	Used as oxygen source in conjunction with fuels or explosives.
Nitroglycerine, NG	H_2C-ONO_2 / $HC-ONO_2$ / H_2C-ONO_2	1.59	2500–7000	Liquid. Very sensitive explosive when solid (m.p. 13.2° C).	Used in DYNAMITES.
Nitrocellulose, NC	$[C_6H_7O_2(ONO_2)_3]_n$	1.45	7010	Sensitive and dangerous explosive.	Used in mixtures with nitroglycerine–GELIGNITE. Commercial and military explosives.

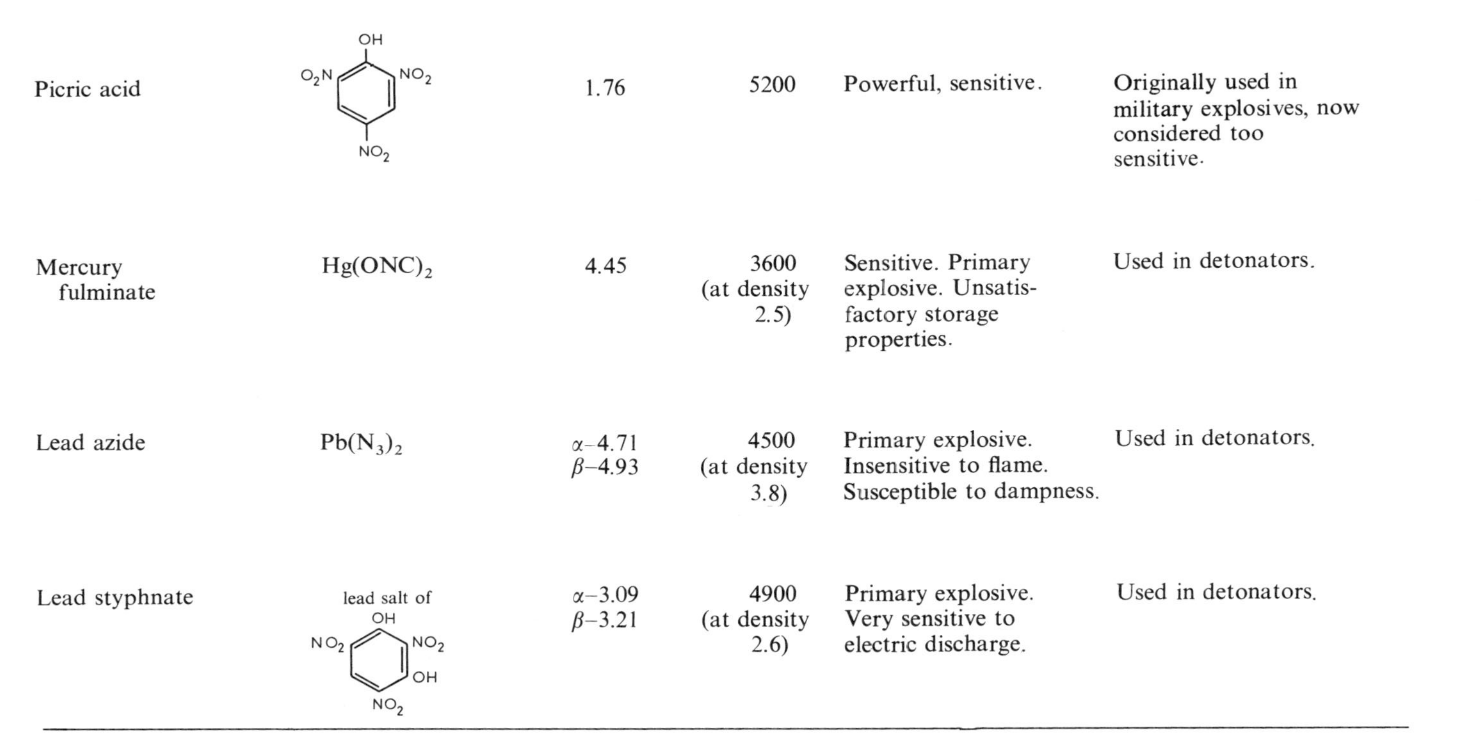

Picric acid	(structure: OH, O_2N, NO_2, NO_2 on benzene ring)	1.76	5200	Powerful, sensitive.	Originally used in military explosives, now considered too sensitive.
Mercury fulminate	$Hg(ONC)_2$	4.45	3600 (at density 2.5)	Sensitive. Primary explosive. Unsatisfactory storage properties.	Used in detonators.
Lead azide	$Pb(N_3)_2$	α–4.71 β–4.93	4500 (at density 3.8)	Primary explosive. Insensitive to flame. Susceptible to dampness.	Used in detonators.
Lead styphnate	lead salt of (structure: OH, NO_2, NO_2, OH, NO_2 on benzene ring)	α–3.09 β–3.21	4900 (at density 2.6)	Primary explosive. Very sensitive to electric discharge.	Used in detonators.

Table 9.2 Composition of some common explosive mixtures (per cent).

Explosive	Composition
Blasting gelignite	Nitroglycerine 91.4, nitrocellulose 8, chalk 0.6.
Ammon gelignite	Nitroglycerine 32.0, ammonium nitrate 60.5, cellulose 6.0, nitrocellulose 1.0, other 0.5
Blackpowder (gunpowder)	Potassium nitrate 75, charcoal 15, sulphur 10
Pentolite	PETN 10–50, TNT 90–50
Amatol	Ammonium nitrate 80, TNT 20
Ammonal	Ammonium nitrate 65, TNT 15, aluminium 17, charcoal 3
Composition A	RDX 91, Beeswax 9
Composition B	RDX 60, TNT 40
Composition C	RDX 88, oil 12
Torpex	RDX 42, TNT 40, aluminium 18

The most important military explosive is trinitrotoluene (TNT). TNT has a low sensitivity and can be melted and cast in shells. TNT produces a high detonation pressure and hence possesses considerable shattering power, or *brisance* (see later), although this can be further improved by the addition of other substances with even greater explosive power, such as PETN or RDX.

A list of typical explosive compositions is given in Table 9.2.

9.3 MILITARY EXPLOSIVES

Although some of the explosives described above find both commercial and military applications, a distinction between the two can usually be made because the requirements of a military explosive are more specific.

Military explosives have to be stored and handled without demanding stringent safety precautions and are therefore subject to considerable hazards of heat and shock prior to use. This constraint is further reinforced by the necessity of being able to fire the shell or missile without causing premature initiation of the explosive charge which it contains. This means that highly sensitive materials cannot be employed in military weapons although they frequently find application commercially, when suitable precautions can be observed.

Military weapons fall basically into three types depending on which of the following effects they are required to produce.

(a) Fragmentation, for example, in the *grenade*, where the intention is to cause injury by flying fragments.

(b) Blast, as in the conventional shell or bomb, to cause structural damage over a wide area.
(c) Pentration, for example, in armour-piercing weapons such as the 'bazooka'.

In all cases, considerable destructive or shattering action, known as brisance, is required, since the bomb or projectile is expected to fragment hard substances such as concrete or steel. Brisance depends on the detonation pressure rather than the explosive strength. This in turn means a high detonation velocity, although the obvious requirement, particularly in a projectile, for maximum destructive power combined with minimum weight necessarily implies high explosive power in addition.

The method of obtaining high penetration involves the *shaped charge* principle described later in the chapter.

9.4 COMMERCIAL EXPLOSIVES

In the majority of commercial operations, a high degree of fragmentation is undesirable, and a high detonation velocity is not required. Instead the explosive may be designed, for example, by suitable selection of grain size, to develop the high pressure over a longer period of time.

Most commercial explosives are made up of mixtures of explosive materials to which non-explosive substances may be added. The first requirement of such a mixture is that it should provide the necessary explosive power. All explosives contain a fuel and an oxidant, although the functions of both may be provided within the same molecule. If the fuel: oxidant ratio does not correspond to the correct stoichiometry, noxious fumes will be produced, an oxygen excess generating nitric oxide and nitrogen dioxide, and an oxygen deficiency giving carbon monoxide. Additional oxygen is usually provided by adding ammonium nitrate. Where oxygen is in excess, the fuel may be made up from a wide range of materials including common organic compounds or even metals. Other compounds may be added to improve the long-term stability of the explosive, to reduce its sensitivity, to provide waterproofing, and to assist in its manufacture, for example, in the form of a plasticizer. The explosives in common use usually fall into one of four categories:

1. Nitroglycerine gelatins ($\rho \approx 1.4$–1.6).
2. Nitroglycerine semi-gelatins and powders ($\rho \approx 0.65$–1.3).

3. TNT powders ($\rho \approx 1$).
4. Ammonium nitrate powders ($\rho \approx 1$).

(Higher density in a particular explosive gives greater explosive power.)

Special explosive mixtures which deserve mention are those used in coal-mining. Mines are very prone to accumulating methane (fire-damp) and if the resulting methane–air mixture is ignited, a disastrous explosion may ensue. For this reason, only *permitted explosives*, which do not ignite such mixtures, are used in underground blasting. In the United Kingdom, permitted explosives are divided into five classes which relate to the nature of the operations for which they are employed. The reason that explosives can be designed which do not cause ignition arises because of the finite induction delay in methane–air mixtures. Explosives are therefore required which have short reaction times and do not produce too high a temperature.

9.5 DETONATORS

The majority of explosives will burn to detonation only if ignited in a suitable manner, otherwise they will simply deflagrate. Ignition is accomplished by a device known as a *detonator* which contains an initiating, or primary, explosive which detonates readily on initiation (Fig. 9.1). Because this material is necessarily very sensitive, it is packed in a metal container to protect it against initiation due to friction or impact. Initiation is then achieved by a flame from a suitable fuse or by electrical means normally using an *exploding bridge-wire*. Detonators may also incorporate a slow-burning material to provide a built-in time delay. These are employed when charges have to be fired separately at predetermined intervals, for example, in tunnelling operations.

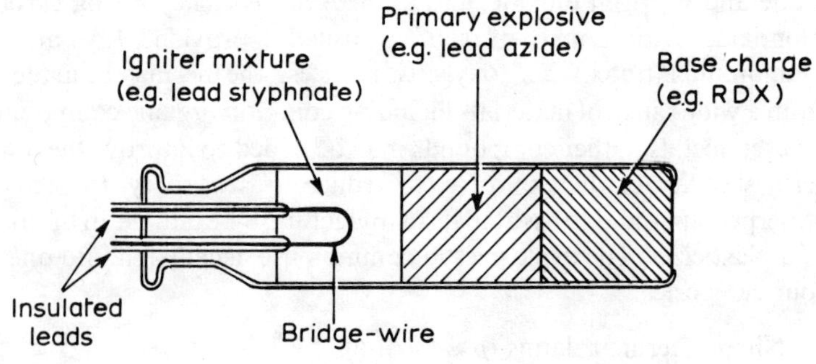

Fig. 9.1 Sketch of an electric detonator.

9.6 BLASTING

The major commercial use of explosives is in blasting operations, notably in mining and quarrying. The technique is basically quite simple. A *borehole* is drilled in the rock to a suitable depth and the explosive charge inserted. A detonator is attached, normally one which can be fired electrically, and the hole is closed by *stemming*.

When the charge is fired, a shock wave is produced in the adjacent rock. The initial force is compressive and causes fracture only in the vicinity of the charge. The shock wave travels outwards and eventually reaches a free surface where it reflects as a rarefaction or tension wave (see Chapter 4). Since the tensile strength of a material is much lower than the compressive strength, the rock will now be considerably fragmented. Depending on the size of the charge and the depth, the total effect will vary between slight surface rupture, through progressively deeper damage and greater fragmentation, to a situation where the rock is broken into quite small pieces and considerable energy is dissipated in throwing the rock out of the crater for appreciable distances and in generating an air-blast. The sequence of effects is illustrated in Fig. 9.2 [171].

The normal requirement is for a *heaving* action, with the rock broken down to the required depth but with the minimum movement of the broken fragments. The operator has to select the correct explosive, charge and depth, in order to achieve the optimum effect. In general, one can say that a hard rock will require an explosive with a high detonation velocity and high energy density whilst for a softer rock it is preferable to select an explosive with a lower detonation velocity in which the pressure build-up is slower. It may be noted that the

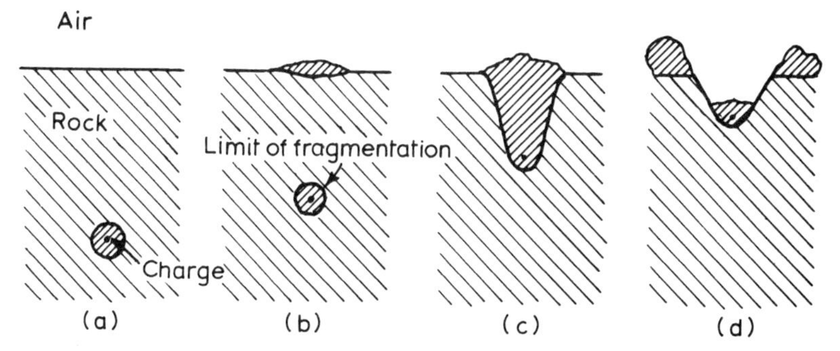

Fig. 9.2 Pictorial representation of the effect of charge depth on the extent of fragmentation (after Fordham [171]).

detonation pressure p_2 and detonation velocity $\boldsymbol{D}$ are related approximately by

$$p_2 = \frac{(\boldsymbol{D}^3 - B\boldsymbol{D}^2)}{4C} \tag{9.1}$$

where the velocity varies with density according to $\boldsymbol{D} = B + C\rho_1$, B and C being constants characteristic of the explosive. The explosion pressure, or initial borehole pressure, is roughly one-half of the detonation pressure. Either the detonation velocity or the explosion pressure may act as guides to the explosive strength obtained. Although it is normally assumed that the explosion process is completed before appreciable work is done on the rock which is to be ruptured, the combustion of many materials occurs by surface erosion and the reaction time can be increased to the order of milliseconds by correct choice of particle size.

Unfortunately the properties of the rock and those of the explosive interact strongly and the two therefore cannot be treated separately. This means that the operator has to fall back on experience, 'rule-of-thumb' or empirical methods, and on-site testing in order to select the necessary conditions. Some indication of the considerations involved is given below.

The succession of events which occur during blasting can be divided into four categories, depending primarily on the depth of the charge. These are:

1. The *strain energy range*, in which some damage occurs in the vicinity of the charge but no surface damage is observed.
2. The *shock range*, in which the tension wave, reflected at the surface, causes the rock progressively to fracture down towards the explosion cavity.
3. The *fragmentation range*.
4. The *air-blast range*, in which the rock is fragmented and excessive energy is dissipated in noise, air-blast, rock movement etc.

The division between (1) and (2) can be specified as the charge depth for which surface damage is first observed and the corresponding depth and weight of explosive are termed the *critical depth*, d_c, and *critical weight*, m_c. The transition between (2) and (3) occurs at maximum efficiency and the corresponding parameters are termed the *optimum depth*, d_{opt}, and *optimum weight*, m_{opt}. Livingston [172, 173] suggested that the properties of a particular combination of explosive and type of rock in the strain energy range can be defined by a *strain energy factor*,

E', given by the relation

$$E' = d_c m_c^{-1/3} \tag{9.2}$$

and that the proportion of the energy which goes into rock fragmentation depends on the *depth ratio*, Δ', which is the depth of the charge, d, divided by d_c. The strain energy factor can be found by determining the value of d_c for a particular weight of charge, and the optimum depth ratio for maximum efficiency by measuring the amount of fragmentation for various charge depths. With these two parameters the correct depth, or *burden*, and charge weight can be estimated for any required fragmentation with that particular rock/explosive combination.

9.7 SHAPED CHARGES

In blasting operations, the charge is usually manufactured in a simple shape for ease of handling. In certain military weapons, particularly those designed for piercing heavy armour, *shaped charges* are employed in which the end opposite to the point of initiation is hollowed out.

Shaped charges make use of the *cavity effect*, also known as the *Munroe effect* or the *Neumann effect*, named after the investigators who discovered it [174–176]. If a charge containing a suitable cavity is fired in contact with a metal plate, an indentation, which is essentially a mirror image of the original cavity, is formed in the plate. This technique is actually used for 'printing' on metal. If the cavity is suitably lined, a deep V-shaped hole is produced. This is because a jet of rapidly moving material is generated due to the collapse of the walls (Fig. 9.3). This feature is employed in weapons such as the bazooka where high penetration is required.

Considerable effort has been devoted to an understanding of the lined-cavity effect because of its obvious importance in military devices. Basically the arrival of the detonation wave at the liner causes it to

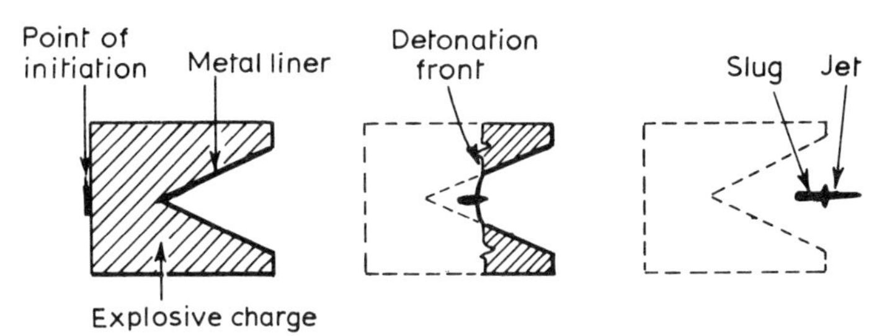

Fig. 9.3 Production of a jet from a lined cavity.

accelerate in towards the cylindrical axis. The convergence of the opposing sides leads to forward and backward velocity components, the former producing the 'jet' and the latter the 'slug'. According to the detonation-head model described in the previous chapter, the products within the head are moving in a forward direction and those behind it are expanding laterally. This means that the forward momentum of the jet will depend on the volume of the detonation head as well as on the detonation pressure. In order to achieve maximum penetration, the geometry of the cavity and liner has to be tailored fairly precisely to the properties of the explosive and a suitable *stand-off* from the target provided to permit complete formation of the jet.

9.8 AIR-BLAST

In addition to the fragmentation and penetration effects achieved in the proximity of an exploding charge, destruction occurs at much greater distances due to the air-blast. The characteristic appearance of blast waves has already been described. As shown in Fig. 4.5, the steep pressure profile is accompanied by an expansion wave and, at appreciable distance from the charge, this wave is primarily responsible for structural damage. In consequence, walls of buildings struck by a blast wave tend to fall out towards the centre of the explosion.

9.9 SUGGESTIONS FOR FURTHER READING

Cook, M.A. (1958) *The Science of High Explosives*, Reinhold, New York.

Fordham, S. (1980) *High Explosives and Propellants*, 2nd edn, Pergamon, Oxford.

Taylor, W. (1959) *Modern Explosives*, Lecture Series No. 5, Royal Institute of Chemistry, London.

10
Rocket propulsion

10.1 INTRODUCTION

The use of combustion systems in jet and rocket motors provides another interesting example of the interplay between chemical and aerodynamic processes. In such motors, the fuel is burnt in a specially designed combustion chamber and the products, normally gaseous, are formed at high temperature and pressure. In the case of a solid fuel rocket, combustion occurs at the exposed surface of the solid, whereas in the liquid fuel rocket, the fuel must first be dispersed in the oxidant and a burner of more complex design is required. A rocket engine does not depend on combustion of a fuel with atmospheric air. Instead, the energy is provided either by the exothermic decomposition of a monopropellant such as hydrogen peroxide or by the combustion of a liquid fuel with an oxidant such as liquid oxygen. Since both fuel and any oxidant are carried on board, a rocket-propelled vehicle can operate outside the earth's atmosphere. The air-breathing jet engine uses atmospheric air as oxidant and the motor is further complicated by the necessity of using a compressor to introduce the air at the correct pressure and velocity.

In all these motors, combustion is performed in a chamber which is open at the rear so that the hot, burnt gases expand backwards through this opening and provide a net force, or thrust, in the forward direction. The conversion of the energy of the products to propulsive effort is therefore a very simple process, requiring no machinery or moving parts. In practice, the rear orifice takes the form of a specially shaped nozzle through which the gases expand adiabatically and which is designed to maximize the thrust obtained (Fig. 10.1).

Once the high-temperature products are formed, the random translational motions of the molecules must be converted to an ordered exhaust velocity. The thrust of the motor, F, is compounded of

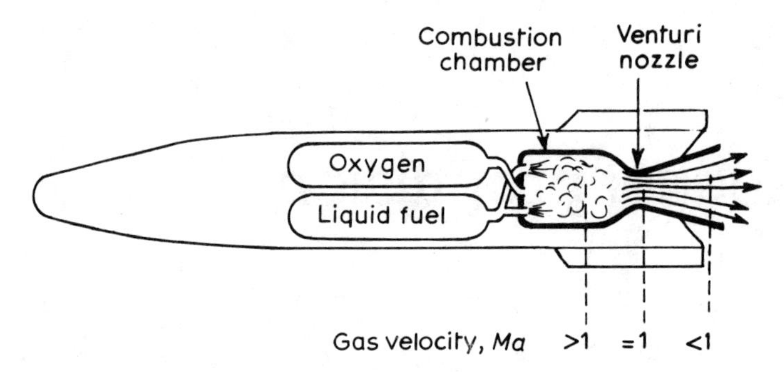

Fig. 10.1 Diagram of a rocket engine.

two terms, the momentum thrust $\dot{m}u_E$ where $\dot{m}$ is the mass produced per unit time and u_E is the exhaust velocity, and the pressure thrust $(p_E - p_0)\,A_E$, where $p_E - p_0$ is the difference between the exhaust and the ambient pressures and A_E is the area of the exhaust

$$F = \dot{m}u_E + (p_E - p_0)A_E \tag{10.1}$$

In alternative designs the exhaust flow is used to drive a turbine connected to a propellor ('turbo-prop') but the considerations remain similar.

10.2 THE EXPANSION NOZZLE

A complication arises in the design of the nozzle because gases in motion behave as incompressible fluids, that is, of constant density, at low velocities but not at high velocities. The continuity equation states that, for steady flow,

$$\rho u A = \text{Constant} \tag{10.2}$$

which is a simple statement of the conservation of mass. This may be written in differential form

$$\frac{d\rho}{\rho} + \frac{du}{u} + \frac{dA}{A} = 0 \tag{10.3}$$

A second equation for fluid motion, which expresses the requirement for conservation of momentum, may be written in the form

$$u\,du + \frac{dp}{\rho} = 0 \tag{10.4}$$

This relationship may be derived, quite simply, from Newton's law, by equating the force acting on a 'particle' of unit volume in steady flow, that is, the pressure gradient $-(\mathrm{d}p/\mathrm{d}x)$, to the product of its density, ρ, and its acceleration $u\,\mathrm{d}u/\mathrm{d}x$. For isentropic flow

$$\frac{\mathrm{d}p}{\mathrm{d}\rho} = a^2 \tag{10.5}$$

where a is the velocity of sound, so that Equation 10.4 may be rewritten

$$u\,\mathrm{d}u + a^2\frac{\mathrm{d}\rho}{\rho} = 0 \tag{10.6}$$

or, introducing the Mach number of the flow $Ma = u/a$ (see Chapter 4)

$$Ma^2\frac{\mathrm{d}u}{u} + \frac{\mathrm{d}\rho}{\rho} = 0 \tag{10.7}$$

Substituting in the continuity Equation 10.3

$$(1 - Ma^2)\frac{\mathrm{d}u}{u} = \frac{-\mathrm{d}A}{A} \tag{10.8}$$

It will be seen that, for subsonic flow with $Ma < 1$, the velocity increases with a reduction in area but, for supersonic flow with $Ma > 1$, the area must be increased to increase the velocity [6]. Thus, a convergent–divergent or de Laval nozzle, as shown in Fig. 10.1, is required to expand the burnt gas to a high (supersonic) exhaust velocity.

10.2.1 Equilibrium and 'frozen' flow

Since combustion reactions are very rapid at high temperatures, the burnt gases will appear in the combustion chamber at local equilibrium and the state of the gas may be obtained from the thermodynamic properties of the initial and final components and the known starting conditions. Because of the high temperatures involved, a high percentage of the product material will exist in dissociated form as atoms and radicals. In order to obtain the maximum flow velocity as the gas expands through the nozzle, equilibrium should be maintained as closely as possible so that, as the temperature falls, enthalpy taken up in dissociation is released as flow energy. Unfortunately, this is difficult to achieve since although the recombination reactions involved usually have a zero or negative temperature coefficient, their high dependence on density means that a stage is reached where the reduction in density through the nozzle is too rapid for equilibrium to be maintained. The flow is then said to be 'frozen' since further expansion occurs without a corresponding shift in the chemical equilibrium. The correct behaviour in the nozzle may be calculated if all the relevant kinetic equations together with the thermodynamic considerations can be taken into account. Even for a relatively simple system, many chemical processes may be involved for which accurate kinetic parameters have not yet been determined. If a complete knowledge of the kinetics were available, lengthy calculations with a high-speed computer would still be required to determine the performance of the system.

Since the calculations are quite straightforward for a fully frozen composition or for complete equilibrium, an estimate of the engine performance may be made if it is assumed that a sharp transition between the two flow regions occurs. A useful criterion suggested by Bray [177] and frequently referred to as the *Bray criterion* states that the flow 'freezes' when the rate of change of the degree of dissociation calculated assuming equilibrium conditions $-(d\alpha_d/dt)_e$ is of the same order as the rate of the dissociation process under the same conditions. In the subsequent discussion, we shall be concerned only with the equilibrium properties of the gas and we must bear in mind that the practical performance will fall somewhat below the theoretical estimate.

10.3 IMPULSE OF ROCKET ENGINES

The performance of a rocket engine is usually expressed by the specific impulse, I_s, which may be defined either as the thrust per unit flow weight of propellant or as the impulse (force × time) per unit weight of propellant consumed. By convention, in engineering practice, specific impulse is reported in seconds.

As the thrust of the engine arises from the conversion of heat energy to flow motion, the specific impulse can be related to the change in enthalpy during the expansion from combustion chamber to exhaust nozzle and, by making the drastic assumptions that the specific heat capacities are independent of temperature and the molecular weight does not change, it can be shown that the specific impulse depends on p_E and p_C, the pressures in the exhaust and combustion chamber and on T_C/M where T_C is the combustion chamber temperature and M the molar mass of the combustion gases.

The values of p_E and p_C depend on the nozzle design, which is primarily an aerodynamic problem, and on the maximum acceptable chamber pressure.

10.4 CHOICE OF PROPELLANT SYSTEM

The chamber temperature, T_C, is determined primarily by the exothermicity of the reaction although, as indicated above, this temperature may be limited by dissociation of products and additional performance is achieved if this energy can be recovered by recombination during the expansion. The effect of M on the specific impulse means that low molecular weight products are to be preferred and this, in turn, puts particular emphasis on hydrogen-containing compounds. Since reactions under these conditions proceed essentially to completion and a series of different reactants can lead to the same products, the kinetics of reaction are of only secondary importance. The performance of propellant systems may then be discussed by considering the reactants and products separately. However, a few comments are valuable regarding the properties of the product gas as a working fluid.

Since the working fluid converts heat energy to flow energy, it is essential that the exhaust gases should have a low enthalpy and therefore a low heat capacity. It is also advantageous to have a large volume of fluid available for expansion. Both these requirements suggest simple, light molecules and weigh against the production of condensed phases which automatically possess high heat capacities and do not contribute to the working fluid. Clearly hydrogen forms a very desirable working fluid and most practical propellants lead to hydrogen in the product gas. The low heat capacity of hydrogen is due in part to its high dissociation energy. A product such as fluorine is less desirable in the working fluid because of its lower dissociation energy although the increase in number of moles on dissociation partially offsets this disadvantage. The major consideration, however, is still that of molecular weight, and the cyanogen–oxygen system with its extremely high flame temperature (~ 5000 K), proves virtually valueless because of the high molecular weight of the products carbon monoxide and nitrogen.

Turning to a consideration of the thermochemistry of propellant systems, it is clear that the products should, for preference, be materials possessing low or negative enthalpies of formation. The specific exothermicity is defined as minus the enthalpy of formation per unit mass of substance and Table 10.1 shows that oxides and fluorides of the lighter elements are potentially useful products but that chlorides, nitrides and carbides are less suitable. Some care is needed in making generalizations of this sort: for example, it is only the gas which behaves as the working fluid and solid products are therefore

Table 10.1 Specific exothermicities of product molecules (MJ kg^{-1}) [178].

Product	Specific exothermicity	Product	Specific exothermocity
[H_2 (g)	0]	HCl (g)	2.6
H_2O (g)	13.4	LiCl (s)	9.6
Li_2O (s)	20.1	$BeCL_2$(s)	5.9
BeO (s)	23.8	BCl_3 (s)	3.2
B_2O_3 (s)	18.4	CCl_4 (g)	0.7
CO (g)	3.9		
CO_2 (g)	8.8	NH_3 (g)	2.7
		Li_3N (s)	5.4
HF (g)	13.4	Be_3N_2 (s)	10.5
LiF (s)	23.4	BN (s)	5.4
BeF_2 (s)	21.3	C_2N_2 (g)	– 5.9
BF_3 (s)	14.2		
CF_4 (g)	10.5	C_2H_2 (g)	– 8.8
		CH_4 (g)	4.6
		Li_2C_2 (s)	– 151

undesirable. Another point worth noting is the importance of boron nitride as a product. Boron nitride is a highly exothermic compound whose effectiveness is reduced by its molecular weight. Hydrogen, on the other hand, is less satisfactory on grounds of exothermicity although very desirable for the other reasons described above. A combination of BN(s) and H_2(g) thus forms a very useful product system and reactants containing B, N and H form valuable propellants. Combinations involving C, N and O also lead to products CO_2 and N_2 with high exothermicities and relatively high molecular weights but, in this instance, the presence of hydrogen is of little assistance because the water-gas equilibrium

$$CO_2 + H_2 \rightleftharpoons CO + H_2O \qquad (10.9)$$

leads to water in the products. In general, oxides and fluorides tend to be the most promising combustion products on grounds of specific exothermicity.

Considering now the reactants, it is evident by corresponding arguments to those above that high endothermicity (high positive enthalpy of formation) is a desirable feature. However, since endothermicity is necessarily associated with high reactivity, the lack of chemical stability of such reactants is likely to prove the practical limitation. The reactants must be chosen not only on the basis of their endothermicity but also with regard to the formation of desirable product species. The oxides and fluorides of the light elements were found most valuable on exothermicity grounds and hydrogen proved to be the most useful working fluid. Hence the hydrides of the light elements are the most common fuels. The ratio of the various elements is also important and may override the simple endothermicity effect. Thus C_2H_4 is a better reactant than C_2H_2 with an F_2 oxidant despite its lower endothermicity. It also follows that oxidants usually contain oxygen or fluorine atoms bonded to light elements, normally other electronegative species.

These general considerations illustrate how various chemical properties can be correlated with propellant performance. Discussion of specific examples will be deferred to a later section.

10.4.1 Engineering considerations

In rocket engines, engineering considerations frequently outweigh considerations based entirely on thrust since the rocket has to support the fuel which it consumes. It is therefore worthwhile listing some of the external factors which have to be taken into account in assessing the merits of a particular fuel.

(a) High density of fuel minimizes weight of storage tanks.
(b) Fuels are most conveniently handled as liquids. Gases which cannot be liquefied at room temperature require refrigeration and hence add to the overall weight of the propulsion system.
(c) High vapour pressure fuels give additional difficulties because the storage tanks must be proportionately stronger and the pumps have to operate at higher pressures to avoid *cavitation* (the formation of bubbles in the fluid when the static pressure falls below the vapour pressure).
(d) Fuels which are stable at room temperature, non-corrosive and non-toxic simplify storage and handling problems.
(e) Fuels are commonly used to cool the combustion chamber, hence they must be thermally stable and preferably have high specific and latent heats, high thermal conductivity and low viscosity.
(f) In cases where intermittent operation is required, a mixture which is spontaneously flammable (hypergolic) makes for easier ignition.

Fuel handling and storage problems are much less critical for solid fuel systems and the additional weight which has to be supported is thereby lower, making solid fuel rockets valuable despite their lower performance. Cooling of the nozzle may become more difficult, however, because *regenerative cooling* is no longer possible. The rheological properties of the 'grain' are important since it essentially forms part of the combustion chamber and is subject to severe stresses particularly during ignition. Thermal expansion properties are also important for the same reason.

This summary of the more obvious design problems illustrates the factors which affect the choice of a propellant for a given task.

10.5 CLASSIFICATION OF PROPELLANTS

Propulsion systems are usually divided into *monopropellants* and *bipropellants*. A monopropellant is a single fluid capable of undergoing exothermic reaction to give gaseous products. Since the system must be stable under storage conditions but react rapidly when ignited, the range of monopropellants is small and largely restricted to solid materials. Monopropellants may be further subdivided into three classes:

Class A. Oxidant and fuel components exist in the same molecule. Most liquid explosives fall in this category but only a small number are effective propellants. Examples are nitroglycerine, nitromethane, nitrate esters, ethylene oxide and hydrogen peroxide.

Class B. Highly unstable compounds. The concept of separate fuel and oxidant functions is of little value, for example, hydrazine, acetylene, propyne.

Class C. A mixture of separate oxidant and fuel constituents. Examples: NH_4OH–NH_4NO_3, CH_3NO_2–CH_3OH, double-base solid propellants (see later).

Bipropellants are stored separately and are only brought together in the combustion chamber. The constituents may therefore react spontaneously and ignition and control of such systems tends to be rather simpler. The constituents are normally classified as oxidants and fuels. Examples are discussed in greater detail below.

10.6 PERFORMANCE OF TYPICAL LIQUID PROPELLANTS

Having examined the general aspects of propellant systems we can now discuss some typical examples in greater detail. Since the number of

Table 10.2 Properties of liquid propellant systems (after Siegel and Schieler [179]).

Fuel	Oxidant	Oxidant: fuel ratio	I_s(s)	Endothermicity of fuel $\Delta_f H^{\ominus}_{298}$(kJ mol^{-1})	T_C(K)	T_E(K)	M_E	Products at T_E (> 5% of total)
H_2	O_2	4.00	391	0	2980	1350	10.08	H_2O, H_2
B_2H_6	O_2	2.00	344	31.4	3846	2592	20.32	$H_2, HBO_2 B_2O_3$
C_2H_2	O_2	1.80	327	227	4172	2600	24.07	CO, H_2O, H_2, CO_2
N_2H_4	O_2	0.90	313	95	3410	1928	20.29	H_2O, N_2, H_2
Kerosene	O_2	2.60	301	− 24.7	3623	2228	25.29	H_2O, CO, CO_2, H_2
H_2	F_2	9.00	410	0	4117	2018	13.64	HF, H_2
B_2H_6	F_2	5.20	371	31.4	4934	3130	22.52	HF, BF, H_2, H
C_2H_2	F_2	1.50	306	227	4109	3294	29.96	HF, C(s)
N_2H_4	F_2	2.30	363	95	4687	2702	21.33	HF, N_2, H_2
Kerosene	F_2	2.40	317	− 24.7	3917	2748	25.26	HF, C(s)
H_2	F_2	9.00	410	0	4117	2018	13.64	HF, H_2
H_2	O_3	3.50	424	142.7	3123	1426	9.07	H_2, H_2O
H_2	F_2O	5.90	410	33.5	3589	1662	11.36	H_2, HF, H_2O
H_2	NF_3	13.30	350	− 124.3	3868	1682	16.43	HF, H_2, N_2
H_2	ClF_3	11.50	318	− 163	3390	1356	16.78	HF, H_2, HCl
H_2	F_2O_2	5.00	407	19.7	3362	1504	10.57	H_2, HF, H_2O

suitable oxidants is smaller than the fuels available it is convenient to classify systems in terms of the former.

(a) H_2–O_2

This combination leads to the highest specific impulse of a liquid oxygen system (Table 10.2) because of the low molecular weight of the products. The optimum mixture ratio is well on the fuel-rich side of stoichiometric simply because the molecular weight of the exhaust gases is lower and compensates for the weight of the additional fuel.

(b) B_2H_6–O_2

Although more exothermic than H_2–O_2, the specific impulse of the system is less because of the higher molecular weight of the products. In this case, of course, additional fuel is less effective in reducing the mean molecular weight. Because of the greater exothermicity, the combustion chamber temperature will be higher and the problem of 'freezing' in the nozzle may prove more serious.

(c) C_2H_2–O_2, N_2H_4–O_2

The performance of both these systems falls a little below that of the diborane–oxygen system. Acetylene fails somewhat as a fuel because, although highly endothermic, the low hydrogen content has an adverse effect on the molecular weight of the exhaust gases. Hydrazine, on the other hand, gains in terms of molecular weight because a major product is water, but loses because of its lower endothermicity.

(d) Kerosene–O_2

Kerosene–oxygen systems are popular on account of the lower cost and ease of handling of the fuel but the specific impulse is comparatively low because of the higher molecular weight of the products formed.

With fluorine as oxidizer, it becomes evident that the greater exothermicity of the system, which leads to greater values of T_C, more than compensates for the increase in molecular weight. This effect is apparent with all the fuels quoted above except acetylene which is less efficient in combination with fluorine. This is a further illustration of the effect of the low hydrogen content of acetylene. Because of the different stoichiometries of

$$H_2 + F_2 \rightarrow 2HF \qquad (10.10)$$

and

$$2H_2 + O_2 \rightarrow 2H_2O \tag{10.11}$$

a higher oxidant ratio is necessary. In general, the production of HF is responsible for the higher thrust obtainable from fluorine-containing systems.

The H_2–F_2 system itself is of particular interest because it is almost the only propellant system for which the reaction mechanism is fully understood. The reaction is not spontaneous and the ignition device presumably serves to produce an initial quantity of atoms. The following sequence then ensues

$$H + F_2 \rightarrow HF + F \tag{10.12}$$

$$F + H_2 \rightarrow HF + H \tag{10.13}$$

$$H + H + M \rightarrow H_2 + M \tag{10.14}$$

$$F + F + M \rightarrow F_2 + M \tag{10.15}$$

$$H + F + M \rightarrow HF + M \tag{10.16}$$

It is possible to calculate the concentration profiles through the expansion nozzle by assuming suitable rate constants for the various reactions and thereby to show that equilibrium is fairly closely maintained throughout.

The other important *cryogenic oxidants* are F_2O, F_2O_2, NF_3, ClF_3, ClO_3F, O_3 and N_2F_4: the relative efficiencies of some of these oxidants with hydrogen as fuel are compared with those of oxygen and fluorine in Table 10.2. Only ozone provides a significant improvement over fluorine, due to its greater endothermicity and the fact that a smaller oxidant:fuel ratio is permitted. The dangers of liquid ozone have, in fact, precluded its use in practical systems. ClF_3 is more attractive than appears from the table because its greater density leads to a higher density specific impulse (see below).

10.6.1 Density specific impulse

The assessment of fuel performance above has been based strictly on specific impulse. Since a rocket is normally required to lift its own fuel, a low-density fuel necessarily adds to the container weight required, hydrogen being particularly unfortunate from this aspect. In order to allow for this effect quantitatively it is convenient to make use of the *density specific impulse*, defined by the relationship $I_{s,\rho} = I_s\rho$. The

Table 10.3 Specific impulse and density specific impulse for some typical propellant systems [177].

Fuel	Oxidant	I_s(s)	$I_{s,\rho}$(g s cm^{-3})
H_2	O_2	391	103
H_2	F_2	410	192
H_2	F_2O	410	158
H_2	ClF_3	318	194
N_2H_4	O_2	313	335
N_2H_4	F_2	363	476
N_2H_4	ClF_3	292	194
Kerosene	O_2	301	322
Kerosene	F_2	317	406
Kerosene	ClF_3	256	356

Note: The ranking order, but not the numerical values, of $I_{s,\rho}$ and I_s may be compared.

values of $I_{s,\rho}$ and I_s for some typical fuel combinations are listed in Table 10.3. It should be noted that relative orders only are meaningful and not the absolute magnitudes. Clearly, hydrogen becomes a much less desirable fuel and ClF_3 is comparable with fluorine as an oxidant on this basis.

An alternative, and preferable, method of evaluating propellant systems is to compare the *terminal velocities* which can be obtained for a particular rocket vehicle. On this basis, as before, the fluorine systems provide the best performance but the H_2–F_2 combination, using a very high oxidant: fuel ratio, is almost as efficient as N_2H_4–F_2. The density specific impulse is therefore only a very crude measure of propellant performance.

10.7 SOLID PROPELLANTS

Some solid fuel motors are based on a monopropellant system in which one chemical compound serves the functions of both fuel and oxidant, for example, nitrate esters such as nitroglycerine or nitrocellulose. In general, mixtures of chemical compounds are used to obtain the required combustion characteristics combined with the desired mechanical properties. Two classes of propellants may be distinguished. *Double-base propellants*, based mainly on a mixture of nitroglycerine, which acts as an explosive plasticizer, and nitrocellulose, which serves as the polymeric binder, take the form of a rigid gel with an essentially homogeneous structure. *Composite propellants* contain the oxidant as finely divided particles dispersed in a polymeric matrix which acts as the fuel. The oxidant is typically ammonium perchlorate, ammonium nitrate, potassium perchlorate or potassium nitrate. Almost all synthetic organic polymers, for example, polysulphides, polyurethanes, vinyl polymers, may serve as the fuel matrix to which fillers such as aluminium metal or carbon black are frequently added. Composite propellants, by virtue of their inhomogeneous structure, show different burning characteristics to the double-base propellants.

The most common oxidant used in solid composite propellants is ammonium perchlorate and with a polyethylene binder the specific impulse reaches a maximum at 90% oxidant of 252 s. Incorporation of aluminium or beryllium leads to a moderate improvement, a mixture of 72% oxidizer, 8% binder and 20% aluminium giving an impulse of 266 s or with 20% beryllium a specific impulse of 280 s. Double-base propellants produce very similar values of the specific impulse. It is clear that solid fuel propellants cannot compete with liquid propellants on the basis of thrust alone.

The burning mechanism of double-base propellants has been described briefly in Chapter 8. Composite propellants differ from double-base propellants in that solid-phase or liquid-phase reactions are relatively unimportant, 'contact' between fuel and oxidant only being achieved when one or both enter the gas phase. Because of the heterogeneous nature of the propellant, separate streams of fuel and oxygen, together with solid particles, are ejected from the surface. The distance required for mixing of the streams depends critically on both the particle size and the pressure, and determines the rate at which heat is transferred back to the burning surface.

10.7.1 Burning characteristics

The pressure dependence of the burning rate of a propellant has very important practical implications. Since airborne missiles have to operate over a wide range of altitudes, the performance must be designed to cope with the variation in barometric pressure.

The burning rates, W, of all propellants show a strong pressure effect which can be described over a reasonable range by the simple empirical relation

$$W = Bp^i \tag{10.17}$$

known as Vieille's law. This relation applies at high pressures, but at other pressures the dependence becomes very complex and, at the lowest pressures, may vanish entirely. The behaviour can be explained qualitatively in terms of the three reaction zones mentioned earlier (see p. 191). The burning rate depends among other things on the rate of surface decomposition and, at the higher pressures, the energy responsible for this decomposition comes from the flame zone. The rate of energy transfer depends on the separation of the flame from the burning surface and on the thermal conductivity. As the flame reaction will be pressure dependent, so the energy transfer and hence the rate of surface decomposition will increase with pressure. In the intermediate pressure range, the thermal energy comes mainly from the fizz zone. This intermediate stage of burning is more complex, depending on specific concentrations of various intermediates somewhat akin to two-stage hydrocarbon combustion. Hence a complex pressure dependence and a strong correlation with composition is to be expected. At very low pressures, energy transfer from the gas phase becomes negligible and the rate shows virtually no pressure dependence, the energy coming entirely from exothermic reaction in the solid.

So far, only steady, one-dimensional burning has been discussed since this is the type of burning aimed for in a rocket motor. However, a number of phenomena which arise from unstable burning and are related to these pressure effects are frequently observed. For example, a small drop in pressure will reduce the rate of the flame reactions and burning may appear to cease. The failure of the surface to decompose will not prevent subsurface reactions from proceeding as before. The evolution of gas will cause a build-up in pressure adjacent to the surface and spontaneous ignition will occur. In consequence, periodic reaction or 'chuffing' can take place. A similar effect may occur with increase in pressure: in this case, standing pressure or acoustic waves can build up and cause resonance burning.

10.8 IGNITION DELAYS

So far, we have been concerned only with the steady-state characteristics of propellant combustion. However, in practice, jet and rocket motors have to be ignited easily and, for many applications, with a very short, reproducible delay. Non-spontaneous propellants naturally require an ancillary ignition device but even with spontaneous propellants an appreciable time may elapse before steady-state burning is established. If unburnt fuel is allowed to build up in the combustion chamber severe pressure transients may occur and a 'hard start' results.

The processes which govern the ignition characteristics are both physical and chemical in origin, delays being associated with vaporization and mixing of liquid droplets, build-up of steady-state concentrations of reactive intermediates, back-diffusion of intermediates and products, and heat transfer from the burning zone. It is therefore necessary to design the combustion chamber and injection systems, the temperature and pressure of operation etc. to minimize the delays produced by these various processes. Because of the complex interplay of different phenomena it is not possible to predict, *a priori*, the absolute delays involved and they must therefore be determined experimentally under conditions which simulate those in the motor as closely as possible.

10.9 SUGGESTIONS FOR FURTHER READING

Altmann, D. and Penner, S.S. (1956) Combustion of liquid propellants, in *Combustion Processes* (eds B. Lewis, R.N. Pease and H.S. Taylor), Oxford University Press, Oxford, p. 470.

Huggett, C. (1956) Combustion of solid propellants, in *Combustion Processes* (eds B. Lewis, R.N. Pease and H.S. Taylor), Oxford University Press, Oxford, p. 514.

Siegel, B. and Schieler, B. (1964) *Energetics of Propellant Chemistry*, Wiley, New York.

11

Internal combustion engines

Internal combustion engines may be subdivided into three main types depending on whether they deliver their energy via a reciprocating piston, a rotating turbine, or direct thrust. The majority of piston engines operate either on the *Otto cycle*, in which the combustion is spark-initiated, or on the *Diesel cycle*, which employs compression ignition. The other engines are based on continuous combustion and do not involve cyclical systems in the same sense. The rocket engine, which produces direct thrust, has been discussed in Chapter 10 and will not be dealt with further. The combustion processes involved in the gas turbine, the turbo-jet and the ram-jet engines, are basically identical and will be discussed altogether.

11.1 THE SPARK-IGNITION ENGINE

11.1.1 Principles of operation

In the conventional piston engine, a fuel–air mixture is introduced into a cylinder, closed at one end, which contains a movable piston. The mixture is compressed by the motion of the piston and is then ignited. The hot, burnt gases at high pressure provide the energy which drives the piston back down the cylinder, the linear motion being converted to a rotary action by a crankshaft and flywheel. A system of valves is arranged so that subsequent movements of the piston can be used to remove the burnt gas and to introduce and compress fresh reactants. Most engines operate on a *four-stroke* cycle, meaning that combustion occurs on one of every four strokes, although some engines are designed for *two-stroke* operation.

Piston engines normally operate at between 500 and 5000 rpm, so that a separate combustion takes place in each cylinder from five to fifty times a second and the motion of each piston is reversed at four times this frequency. The shape of the region of the cylinder within which combustion occurs and the design of the valve gear are therefore critical in obtaining efficient charging and discharging of the gases in these short time intervals.

The construction of a typical piston engine is shown schematically in Fig. 11.1. On the four-stroke Otto cycle, operation commences with a downward motion of the piston with the inlet valve open. During this *induction* stroke, the correct fuel–air mixture is admitted to the cylinder. The valve closes on the upward, or *compression*, stroke and the mixture is compressed and heated adiabatically. Shortly before *top dead centre*, the highest point of piston travel, the mixture is ignited by a high-tension spark and combustion proceeds essentially to completion

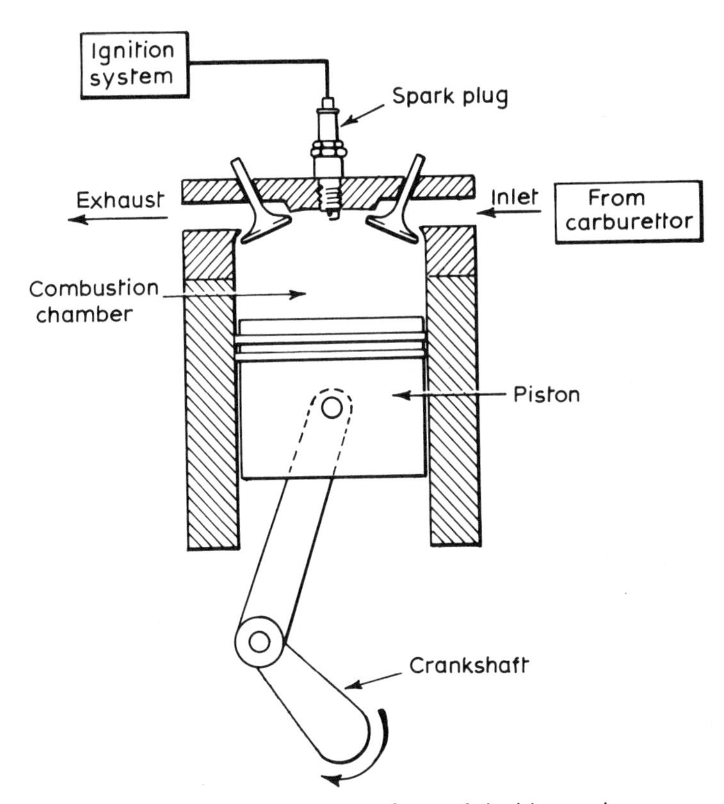

Fig. 11.1 Schematic diagram of a spark-ignition engine.

before the piston has had time to move appreciably downwards. Combustion therefore occurs essentially at constant volume. The hot, high-pressure gas then drives the piston downward on the *power* stroke. On the final upward, or *exhaust*, stroke of the cycle, the exhaust valve opens to enable the burnt gases to leave. Both valves normally open before the stroke commences and close after it has finished to take maximum advantage of the momentum of the gas flow in transferring it to and from the cylinder.

The fuel–air mixture is obtained by passing air through a *carburettor* in which fuel is drawn into the flow as a fine spray by making use of the Venturi effect, that is, the reduction in pressure which occurs when the velocity of a fluid increases on passing through a constriction. This unit is quite complex since it must permit a wide variation in throughput, usually over a factor of ten, and also alter the mixture strength with variation of engine load. The carburettors and inlet manifolds employed in automobile engines are relatively inefficient at delivering the correct mixture to each cylinder. This is a prime source of the air pollution generated by such engines and has led to the wider adoption of direct *fuel injection* which is far more efficient in this respect, although more expensive.

The timing of the spark is quite critical and can usually be adjusted, in principle, to within 20 μs. Furthermore, this timing is required to change both with engine speed and with engine load. The performance of an engine of this type and, in particular, its reliability depend very much on these ancillary functions, carburation and ignition.

11.1.2 Performance considerations

The behaviour of an internal combustion engine may be recorded on an *indicator diagram* which shows the cycle as a p–V plot. A typical diagram is illustrated in Fig. 11.2(a). The point A represents the situation at top dead centre immediately prior to commencement of the induction stroke. At this stage, the space above the piston contains some burnt gas left over from the previous cycle. AB then represents the induction stroke and BC the compression stroke returning the piston to top dead centre. The mixture is ignited at X, shortly before this point. CDE forms the power stroke and EA the final exhaust stroke which completes the cycle.

The total area $\int p\mathrm{d}V$ measures the work done by the engine, the hatched area being negative in the sense that it represents the losses involved in moving gases in and out during the inlet and exhaust

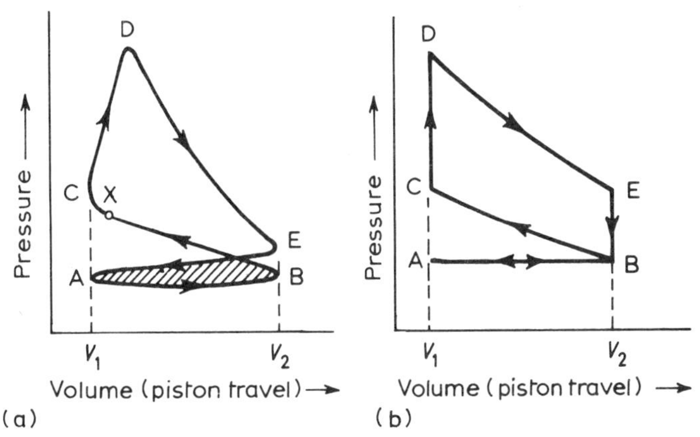

Fig. 11.2 Indicator diagrams for a spark-ignition engine. (a) Typical diagram for a practical engine. (b) Diagram for an ideal air Otto cycle.

strokes, respectively. The theoretical energy available can be calculated from the heat of combustion of the fuel and the mixture composition so that the experimental indicator diagram may be used to calculate the amount of energy converted to work, that is, the thermal efficiency of the engine.

In order to understand the operation further it is preferable to consider the idealized cycle shown in Fig. 11.2(b). The pressure changes consequent on combustion and on opening and closing the valves are assumed to occur instantaneously so that the lower loss area is non-existent. (Note that the exhaust stroke becomes EBA and the induction stroke AB.) Furthermore, the gas is assumed to behave ideally throughout, with a constant specific heat and no mole number change. This is not in fact as bad an approximation as it appears since the major component of the actual mixture is nitrogen, whose specific heat varies little with temperature.

It is then quite simple to show that the heat produced during combustion per unit mass of gas must equal $c_V(T_D - T_C)$ and that the heat lost due to exhausting hot burnt gas and drawing in fresh mixture is $c_V(T_E - T_B)$. Thus the thermal efficiency, ε, of the engine is simply

$$\varepsilon = \frac{(T_D - T_C) - (T_E - T_B)}{(T_D - T_C)} = 1 - \frac{T_E - T_B}{T_D - T_C} \tag{11.1}$$

If the compression and power strokes occur isentropically then

$$\frac{T_B}{T_C} = \frac{T_E}{T_D} = \left(\frac{V_1}{V_2}\right)^{\gamma - 1} \tag{11.2}$$

where $\gamma = c_p/c_V$. Substituting in Equation 11.1 gives

$$\varepsilon = 1 - \left(\frac{V_1}{V_2}\right)^{\gamma - 1} = 1 - \frac{1}{(\text{c.r.})^{\gamma - 1}} \tag{11.3}$$

where $V_2/V_1 = \text{c.r.}$, the *compression ratio* of the engine. Thus the maximum efficiency of the engine is obtained by working at high compression ratios and high values of the specific heat ratio γ; since air has a higher γ value than the fuel, the latter condition corresponds to a requirement for weak mixtures.

The work done per cycle is given by the product of the chemical energy released and the thermal efficiency. For the ideal cycle, the former is simply the amount of fuel (or mixture) admitted, measured in moles, multiplied by the heat of reaction at constant volume, $-\Delta U$. From the ideal gas law the amount of gas admitted is $(p_0/RT_0)(V_2 - V_1)$ where the subscript 0 refers to conditions at the inlet manifold. The work done on the piston is therefore equal to

$$\varepsilon(-\Delta U)(p_0/RT_0)(V_2 - V_1)$$

Since the energy is produced by moving a piston through the fixed volume, $V_2 - V_1$, it is convenient to express the performance of the engine by the mean effective pressure which equals $\varepsilon(-\Delta U)p_0/RT_0$. The work done by a real engine is given by the area of the indicator diagram, and thus the mean effective pressure (mep) is

$$\text{mep} = \frac{\text{Area of indicator diagram}}{\text{Volume displaced}} \tag{11.4}$$

Some of the chemical energy released by reaction may be lost in dissociation at high temperatures so that the net useful energy released reaches a maximum with mixtures which are richer than stoichiometric, although the maximum temperature is slightly lower. Burning velocity is also a maximum with mixtures somewhat on the rich side of stoichiometric. As a result, although rich mixtures give maximum power, maximum economy is obtained with weak mixtures. These factors have important consequences for carburettor design. Conventional internal combustion engines have hitherto almost always operated with slightly rich mixtures for best performance, and also because the products resulting from weak mixtures cause damage to the material from which engines are constructed, but recent emphasis on fuel economy is tending to change this practice.

For a given fuel, the power output will be increased by lowering T_0 and raising p_0. Little reduction can be effected in T_0 with normal carburettors because of the necessity of vaporizing the fuel. On the other hand, p_0 may be increased artificially by *supercharging*. Supercharging is essential for aircraft engines to prevent severe power losses at high altitude and, in various forms, is applied to other engines too.

The so-called *pumping losses* neglected in the idealized cycle are clearly important factors in engine design. The actual combustion process is closely associated with turbulence in the combustion chamber and this again depends on the motion of gas in and out of the cylinder during the induction and exhaust strokes. Increasing the turbulence enhances the burning velocity and hence the rate of combustion (see Chapter 3). The whole question of engine 'breathing', whilst very interesting, falls into the realm of aerodynamics and cannot be considered further here.

11.1.3 Knock in engines

It will be noticed that both the efficiency and the power of an engine increase with compression ratio and it would seem advisable to design engines with the highest possible value of this ratio. Unfortunately, as the compression ratio is raised a phenomenon known as engine *knock* is experienced. When the mixture is ignited by a spark, the initial combustion wave is that of a pre-mixed flame but in addition the gas ahead of the flame front will be compressed and heated. If this is sufficient to induce spontaneous ignition, then *knocking combustion* occurs. This causes irregular burning which interferes with the smooth motion of the piston and hence reduces the power and makes the unit noisy. Fig. 11.3 illustrates the difference between the smooth pressure change in the cylinder during normal combustion and the 'spiky' pressure change seen when knocking is taking place. Knock also increases the turbulence in the combustion chamber and may allow more rapid transfer of heat to the walls. The destructive effect can have very serious consequences, particularly in supercharged engines where the running temperature tends to be higher. These factors set an upper limit on the compression ratio of engines designed for normal commercial operation of between 9:1 and 10:1 with the fuels currently available although with weak mixtures, higher compression ratios are possible.

The terminology associated with engine knock occasionally causes confusion. *Pre-ignition* refers to the ignition of the reactants in a spark-ignition engine prior to firing of the spark and usually occurs on overheated surfaces which stay hot throughout the cycle. *Auto-ignition*

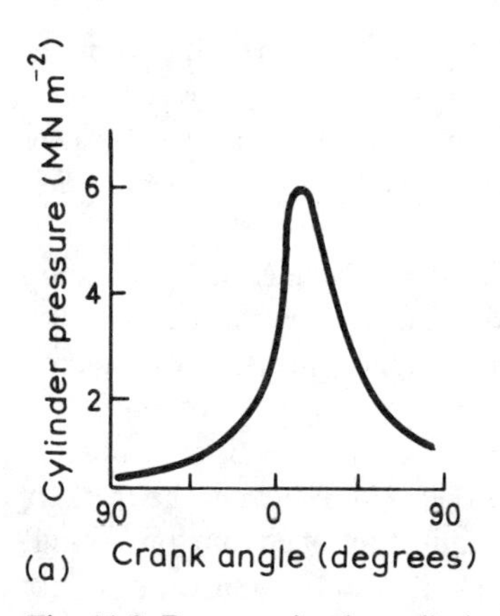

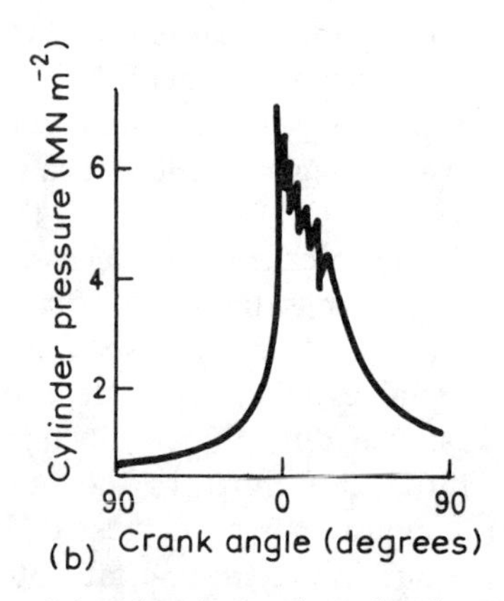

Fig. 11.3 Pressure in the cylinder of a spark-ignition internal combustion engine (a) during normal running, (b) during knocking combustion.

('pinking', 'knocking', 'detonation') denotes spontaneous ignition of the unburned portion of the fuel–air mixture (*end-gas*) due to adiabatic compression; in spark-ignition engines, compression is brought about by the motion of the piston and by expansion of the burned gas behind the moving flame front [180]. Knock and detonation are frequently used synonymously although the effects of auto-ignition are experienced without a true detonation wave necessarily forming.

The preparation of commercial petroleum fuels involves many considerations. These include the insertion of low-boiling constituents for easier vaporization and hence better cold starting without the formation of vapour locks, freedom from high-boiling components which lead to carbon formation, the absence of water and acidic impurities which cause corrosion, and freedom from substances which produce gums, tarry deposits and offensive odours. However, the prime factor still remains the *knock limit* which is determined by the nature of the fuel. Typical compression ratios at which knock occurs for various fuels are listed in Table 11.1 [181].

The knock resistance is usually measured by the *octane number* of the fuel, which is defined as the percentage by volume of *i*-octane, $CH_3C(CH_3)_2CH_2CH(CH_3)CH_3$, in a mixture of *i*-octane and *n*-heptane, which has the same tendency to knock under standard engine conditions as the fuel. As fuels are now in use which have a greater knock resistance than pure *i*-octane, it has proved convenient to extend the octane number scale above 100 using mixtures of *i*-octane and lead tetraethyl. Octane numbers usually increase as the fuel

Table 11.1 Critical compression ratios and octane numbers for some typical hydrocarbons [181] ('Motor' and 'Research' refer to different methods of measuring octane numbers).

Fuel	Formula	Critical compression ratio (at 600 rpm, 177° C)	Octane number	
			Research	Motor
Methane	CH_4	13.0	—	—
Ethane	C_2H_6	9.4	101.7	100.05
Propane	C_3H_8	8.8	101.6	97.1
n-Butane	C_4H_{10}	5.3	94.0	89.1
n-Pentane	C_5H_{12}	3.2	61.8	63.2
2-Methylbutane	$CH_3CH(CH_3)CH_2CH_3$	5.1	92.3	90.3
n-Hexane	C_6H_{14}	3.0	24.8	26.0
2-Methylpentane	$CH_3CH(CH_3)CH_2CH_2CH_3$	3.65	73.4	73.5
2, 2-Dimethylbutane	$CH_3C(CH_3)_2CH_2CH_3$	5.1	91.8	93.4
n-Heptane	C_7H_{16}	3.0 (at 100° C)	0	0
2, 2-Dimethylpentane	$CH_3C(CH_3)_2CH_2CH_2CH_3$	5.55	92.8	95.6
2, 4-Dimethylpentane	$CH_3CH(CH_3)CH_2CH(CH_3)CH_3$	4.35	83.1	83.8
2, 2, 3-Trimethylbutane	$CH_3C(CH_3)_2CH(CH_3)CH_3$	10.5	101.8	100.1
2, 2, 4-Trimethylpentane (*i*-octane)	$CH_3C(CH_3)_2CH_2CH(CH_3)CH_3$	6.6	100	100
Ethene	C_2H_4	5.6	97.3	75.6
Propene	C_3H_6	6.9	100.0	84.9
1-Butene	$CH_2{:}CHCH_2CH_3$	5.3	97.5	79.9
cis-2-Butene	$CH_3CH{:}CHCH_3$	5.95	100	83.5
1-Pentene	$CH_2{:}CHCH_2CH_2CH_3$	4.55	90.9	77.1
2-Methyl-2-butene	$CH_3C(CH_3){:}CHCH_3$	6.45	97.3	84.7

Table 11.1 (*Contd.*)

Fuel	Formula	Critical compression ratio (at 600 rpm, 177° C)	Octane number	
			Research	Motor
Cyclohexane	CH_2 CH_2 CH_2 CH_2 CH_2 CH_2	4.7	83	77.2
Benzene		—	—	102.7
Toluene	CH_3	11.35	105.8	100.3
Ethylbenzene	CH_2CH_3	8.2	100.8	97.9

becomes more difficult to oxidize so that the knock resistance is greater for shorter carbon chains or a higher degree of branching in the fuel. Apart from ethene and propene, the alkenes have a higher knock resistance than the corresponding alkanes. The behaviour of alcohols appears to parallel that of alkenes.

The cause of knock is the occurrence of phenomena such as cool flames and two-stage ignition, and knock is therefore associated with the branching-chain character of the reaction. The onset of knock can frequently be correlated with the induction times for such effects and with the cool flame behaviour of fuels [182].

Furthermore, there are close parallels between the products from cool flames in static systems and from engines, cool flame reactions appearing to reach their peak just before the onset of knock [183].

The generalized model of hydrocarbon combustion outlined in Chapter 7 has been incorporated in an engine cycle simulation to yield a computer model of engine knock [184]. This model confirms the origin of knock as the homogeneous ignition of the gas ahead of the flame front in the cylinder and can also model other combustion processes of practical importance in engines, such as *run-on* and the ignition of fuel sprays.

Certain additives, known as *anti-knocks*, cause a pronounced increase in knock resistance, even when present in only trace amounts. These substances almost certainly act by destroying radicals and hence preventing chain reactions from building up. In general, anti-knocks are organo-metallic compounds and by far the most common are lead tetraethyl, $Pb(C_2H_5)_4$, and lead tetramethyl, $Pb(CH_3)_4$. Even with these materials the detailed mechanism is not understood, mainly because the effect on engine knock cannot easily be related to their effect on cool flames and pre-ignition reactions.

It has been established that lead tetraethyl functions through the elimination of the chemical precursors of the knock reactions and not through any direct influence on the rate of flame propagation [185]. There is general agreement that lead tetraethyl is converted into lead oxide, PbO, in the engine but the way in which this suppresses knock has been a matter of great debate [106, 186]. Although a possible gas-phase mechanism has been suggested [186–188], a 'fog' or colloidal dispersion of solid lead oxide particles about 10 nm in diameter is formed in the cylinder end-gas before the arrival of the flame and the balance of the evidence strongly favours a heterogeneous process in which chain-carrying radicals are removed on the surface of solid lead oxide particles [189–191]. There is further support for the heterogeneous mechanism from a study of the decomposition of lead tetramethyl in a shock tube. This work has shown that particles of lead monoxide could be formed in time to influence pre-ignition reactions [192].

Because lead compounds are toxic, a great deal of effort has been devoted to discovering other anti-knock agents without this undesirable property. The choice of compounds is restricted by practical requirements; for example, an anti-knock must be miscible with the fuel and it must be stable and volatile. Methyl *t*-butyl ether and *t*-butanol are among the compounds which show some promise of being able to replace lead additives as anti-knocks in motor vehicle fuels.

Other additives, known as *pro-knocks*, tend to promote knocking combustion. These compounds also accelerate slow combustion and presumably act by causing chain initiation. Such compounds as aldehydes, organic nitrites and nitrates, and alkyl peroxides, are all effective pro-knocks. They have been used, particularly in wartime, to improve the starting characteristics etc. of low-grade fuels. They are also used as ignition promoters in diesel engines.

If lead additives in petrol are to be reduced or even eliminated, it is probable that engines will have to run on lower octane fuel with consequent reductions in performance and fuel economy. A cognate problem (also discussed in Chapter 13) is the need to reduce the emission of pollutants and many possible solutions have been proposed. One promising technique is water injection. This has a long history and was used in supercharged aircraft engines to prevent damage from severe knock and overheating. It was also employed in early turbo-prop and jet engines to extend the high power limit and there have been a few commercial applications to road vehicles [193, 194]. Recently some turbo-charged racing car engines have employed water injection. The main function of the water is as a coolant and quite a small amount produces a significant improvement in the performance of these high-output engines operating close to their knock limit. Various means have been used to add water vapour to the fuel–air mixture in the inlet manifold and it is clear that water addition enables a spark-ignition engine to run economically with unleaded fuel in a high compression ratio engine; the emission of NO_x, in particular, is markedly reduced. In diesel engines, water injection also has beneficial results on emissions, not only of NO_x but also of soot [195]. The mechanism by which water vapour functions is not clear but the lowering of NO_x in the exhaust is probably associated with lower temperatures in the combustion gases. Although water injection has certain advantages, it has not been widely adopted so far, principally because of the additional complexity of the equipment required. Some of the other ways of reducing emissions are discussed later in this chapter and in Chapter 13.

11.1.4 Lean-burn engines

As Equation 11.3 shows, the efficiency of a four-stroke Otto-cycle engine can be improved by raising its compression ratio; it can also be improved by operating with a lean mixture, that is, one with a higher air:fuel ratio than the stoichiometric ratio of about 15:1 by weight. These changes can also lead to a reduction in the emission of pollutants.

Increasing the compression ratio is limited by the onset of knock and although weaker mixtures allow the ratio to be raised further before auto-ignition sets in, they tend to burn rather more slowly and are prone to misfiring. This can be counteracted, however, by making the fuel–air mixture in the cylinder more turbulent, and high-turbulence lean-burn engines can operate at compression ratios of 13:1 instead of the 9.5:1 currently used with fuels of 96–98 octane.

11.1.5 Stratified-charge engines

Lean mixtures may also be burned successfully in a two-part combustion chamber; a small pre-chamber is supplied with an easily

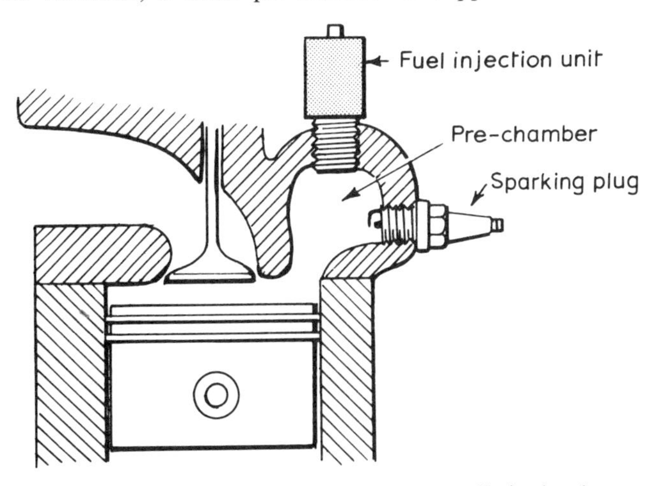

Fig. 11.4 Diagram of a stratified-charge cylinder head.

ignitable rich mixture from which the flame spreads to the main region containing the leaner mixture (Fig. 11.4). Various types of *stratified-charge* engine have been developed, some with fuel injection, which use this principle.

Other versions of the stratified-charge engine use specially shaped combustion chambers and pistons. In one such design a curved channel in the cylinder head leads from the region of the inlet valve to the combustion space which is concentrated under the exhaust valve. As the piston approaches the top of its stroke, turbulence is induced in the flow in this channel (Fig. 11.5).

In engines of these types, mixtures with air:fuel ratios of 18:1 or 20:1 (in contrast to about 15:1 used in normal engines) have been burned successfully with considerable reduction in the levels of pollutants in the exhaust gas combined with improved fuel economy [196].

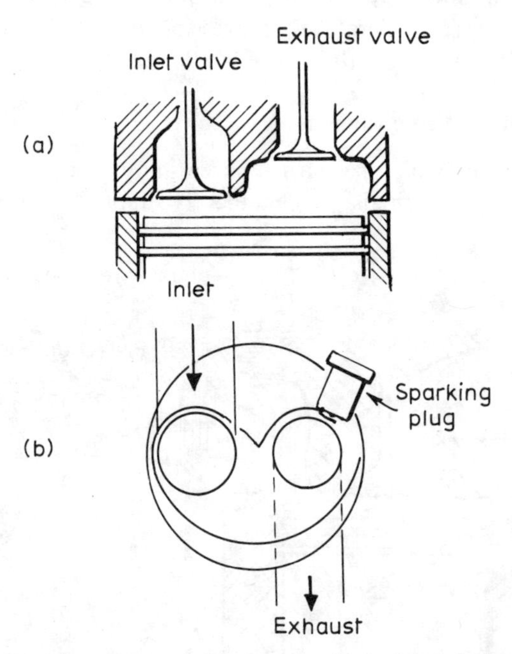

Fig. 11.5 Diagram of a specially shaped cylinder head designed to promote turbulent swirl in the combustion gases. (a) side view, (b) plan view.

11.2 THE DIESEL ENGINE

11.2.1 Principles of operation

The operation of the diesel engine depends on the same four-stroke cycle of a reciprocating piston as the spark-ignition engine, that is, induction, compression, power and exhaust. However, instead of pre-mixing fuel and air prior to induction, air alone is taken into the cylinder and compressed adiabatically. At the top of the compression stroke, liquid fuel is injected into the cylinder and ignites on contact with the hot air. The injection of fuel continues for a substantial portion of the power stroke. The diesel engine is also known as a *compression-ignition engine* and typical compression ratios are in the range 14:1 to 20:1.

In order to achieve spontaneous ignition, the compression stroke must raise the air to a much higher temperature than in the spark-ignition engine and the design of the diesel engine is characterized by a much greater compression ratio. Because the design features are based on an engine which has attained a steady operating temperature, the diesel engine is notoriously difficult to start from cold. However, once in motion, the engine is more rugged and reliable because of the simplicity of the ancillary systems. The spark-ignition engine requires a high-intensity spark on each cycle at a time which depends both on the rate of revolution of the engine and on the power which it develops, that is, on the applied load. The compression-ignition engine needs no ancillary electrical equipment. Whilst conventional spark-ignition engines also require the fuel to be vaporized, mixed with air and the correct quantity admitted via a carburettor, in the diesel engine the flow of liquid fuel is metered directly into the cylinder. Finally, the latter will normally run on far cruder fuel. These engines have considerable advantages in commercial applications, for example, for continuous, heavy-duty operation such as is required in locomotives, heavy road transport and marine engines, or where robustness is at a premium as in tractors and earth-moving vehicles.

11.2.2 Performance considerations

An ideal air cycle for such an engine is shown by the indicator diagram in Fig. 11.6. The main difference from the spark-ignition engine is in the power stroke. Combustion occurs smoothly at *constant pressure* during the first part of the power stroke CD and when complete, the gas

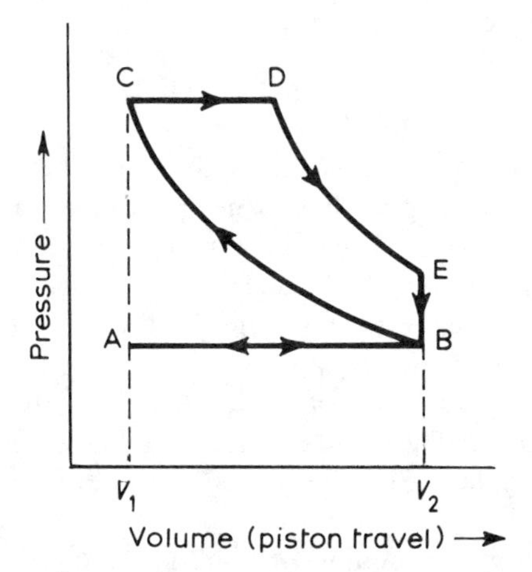

Fig. 11.6 Indicator diagram for an engine operating on an ideal air diesel cycle.

continues to expand along a portion of the power stroke DE. This contrasts with the spark-ignition engine (Fig. 11.2(b)) where the combustion occurs very rapidly and therefore essentially at constant volume.

By a similar calculation to that outlined in the previous section, it can be shown that the thermal efficiency, ε, is given by

$$\varepsilon = 1 - \frac{1}{(\text{c.r.})^{\gamma-1}}\left(1 + \frac{\gamma - 1}{2}x + \cdots\right) \tag{11.5}$$

where

$$x = \frac{V_D}{V_C} - 1 = \frac{(\text{c.r.}) - 1}{(\text{c.r.})^{\gamma}} \frac{(-\Delta H)}{c_p T_0}$$

and is less than unity for efficient operation. Once again the thermal efficiency increases with greater compression ratios and weaker mixtures but, for identical values of the two, the efficiency of the Otto cycle is always greater.

The power output, as indicated by the mean effective pressure, follows the same relationship as in Equation 11.4 and would therefore indicate that maximum power would be available for rich mixtures exactly as in the Otto cycle. This deduction proves to be of academic interest only, since the inefficient mixing of fuel and air means that diesel engines are always operated with weak mixtures to eliminate incomplete combustion of the fuel.

11.2.3 Combustion processes in the diesel engine

The processes occurring in the diesel engine possess the complexity characteristic of heterogeneous combustion. In an idealized situation, the fuel would enter as a fine spray of liquid droplets distributed evenly throughout the combustion chamber. The fuel would then vaporize completely before ignition occurred. In order to achieve this as closely as possible, the fuel is forced in at high pressure and distributed by a series of fine sprays, the air being simultaneously 'swirled' to assist in turbulent mixing.

In that part of the cycle during which fuel is being injected there is an appreciable induction delay, followed by a steep pressure rise and then a rapid combustion region in which the reaction rate is controlled by the rate of fuel injection. The induction delay allows evaporation of some fuel before burning commences. However, the subsequent pressure rise leads to a form of knocking which may interfere with the smooth motion of the piston. Due to the accumulation of unburnt fuel, this knocking becomes more severe the greater the induction delay. The greatest hindrance to satisfactory operation occurs once steady burning has commenced. In this stage of reaction, the incoming fuel has insufficient time to vaporize and mix completely so that local regions of very rich mixture occur and soot tends to form in the gas and pyrolytic residues are deposited directly on the walls. Both will eventually prevent satisfactory engine operation. The latter problem is usually overcome by employing a large excess of air which delays the ignition and permits better mixing although at the expense of power output. Since the various stages require different physical conditions, the optima were determined by trial-and-error methods but this is giving way to more sophisticated techniques including mathematical modelling.

These considerations show that the advantages of the compression-ignition engine of ruggedness and durability are paid for in terms of reduced economy and power output.

For relatively small high-speed diesel engines, as used in passenger cars and light goods vehicles, the only satisfactory way of achieving the short ignition delay and high degree of mixing needed has been by the use of *indirect injection*. High turbulence is generated by forcing about half the air trapped by the piston into a separate pre-combustion chamber. All the fuel is injected into this chamber at a compression ratio of above 20:1. Combustion starts in this rich mixture and spreads into the leaner mixture in the cylinder as the piston descends. There are similarities between this system and the stratified charge principle in spark-ignition engines. Low emissions and noise are obtained with peak pressures not much above those encountered in spark-ignition engines. On the other hand, fuel economy is relatively poor.

Direct injection diesels are normally limited to speeds below about 3000 rpm and compression ratios of around 20:1, these limits being set by the ignition delay time. The peak pressure during the cycle may exceed 100 atm. Fuel consumption is lower and starting easier, but direct injection engines are noisy and suffer from considerable problems with gaseous emissions.

Because ignition occurs by compression and heating, the pre-flame reactions which cause knocking in the spark-ignition engine are precisely those which are required in the diesel engine. This means that the relative merits of different fuels are almost exactly the reverse of their behaviour in Otto-cycle engines and compounds which are pro-knocks in spark-ignition engines function as ignition promoters in diesel engines. The suitability of a diesel fuel is given quantitative significance in the *cetane number*. The behaviour of the fuel is measured

Table 11.2 Cetane numbers of some typical hydrocarbons (after Ham and Smith [197]). Corresponding octane numbers have been included for comparison where available.

		Octane number	
Fuel	Cetane number	Research	Motor
n-Heptane	56.3	0	0
n-Octane	63.8		
n-Decane	76.9		
n-Dodecane	87.6		
n-Tetradecane	96.1		
n-Hexadecane (cetane)	100.0		
Octadecane	102.6		
1-Octene	40.5	28.7	34.7
1-Decene	60.2		
1-Dodecene	71.3		
1-Tetradecene	82.7		
1-Hexadecene	84.2		
1-Octadecene	90.0		
Methylcyclohexane	20.0	74.8	71.1
Dicyclohexyl	47.4		
Decalin	42.1		

in a standard engine test, the cetane number being the percentage of cetane, that is, hexadecane, in a mixture of cetane and α-methyl-naphthalene, which gives the same results. Cetane numbers of some typical fuels are listed in Table 11.2 for comparison with the octane ratings although there is little overlap on the two scales. A good diesel fuel must also be capable of ready spark ignition well below the self-ignition temperature to ensure good starting from cold, and the delay between injection and ignition must be small otherwise rough running will occur.

11.2.4 Alternative fuels for spark-ignition and diesel engines

The increasing cost of crude oil has prompted a search for fuels to replace motor gasoline in spark-ignition engines and diesel oil in compression-ignition engines. The problem is a difficult one because the widespread use of any novel fuel requires the setting up of a new and very costly distribution system. In addition, engine modifications are generally needed if performance and reliability are to be maintained.

The alternative fuel which has attracted most attention is ethanol, largely because it can be derived from natural sources by fermentation. Other possible substitute fuels include liquefied petroleum gas (LPG) which is comprised mainly of propane and butanes, compressed natural gas (CNG), vegetable oils and synthetic fuels derived either from coal or natural gas (methane). In every case, there is a substantial penalty in that for a given distance travelled by the vehicle, the weight of the substitute fuel and its container is greater than that of the comparable amount of gasoline and its tank. LPG and CNG suffer particularly badly in this respect and, in addition, require specialized equipment for distribution and refuelling.

It is technically possible to run existing spark-ignition engines, with only slight modification, on some of these novel fuels and, indeed, methanol and ethanol have good knock-ratings and engines using them run cool and emit relatively small amounts of pollutants; mixtures of alcohol and gasoline have also been used. LNG and CNG are other fuels with good octane numbers and low emission characteristics which can be used in conventional engines. However, the greatest advantages accrue with specially designed high-compression engines.

The current versions of diesel engines cannot run on alcohol fuels and for these, fuel oils derived from vegetable sources such as soya beans, sunflower seeds and coconuts seem promising in those parts of the world where petroleum products are costly or in short supply.

The widespread adoption of new fuels will be influenced by many non-technical factors such as government policy, and it seems likely that the change will generally be slow, except in countries without ready access to oil-based fuels.

11.3 GAS TURBINE AND JET ENGINES

The operating principle of all these engines is similar. The air is first compressed and then fuel is injected into and burnt in the high-pressure air. The burnt gas at high pressure and temperature then provides the energy source. In the *gas turbine*, the whole of this gas is used to drive the turbine whilst in the *turbo-jet* sufficient energy is taken from the exhaust gas only to drive the compressor, the remainder providing the propulsive thrust for the vehicle, normally an aircraft. In the *ram-jet*, the compression is provided entirely by shock waves formed in the intake due to forward motion of the engine. The whole of the exhaust gas is then available for propulsive thrust. The three engines cover the complete 'spectrum' of operation and it is only necessary to illustrate the turbo-jet unit (Fig. 11.7).

The design of a gas turbine engine is a complicated affair because of the need to meet a very wide range of exacting operating conditions. The thermodynamics of the combustion process assumes less importance as it is necessary to keep the temperature of the interior of the motor below a critical value dependent on the constructional material employed; consequently, very weak mixtures are used. A major design problem is thus that of maintaining a stable flame in a high-velocity stream of gas in which the overall fuel:air ratio is below the lean flammability limit. In addition, the engine has to operate at different

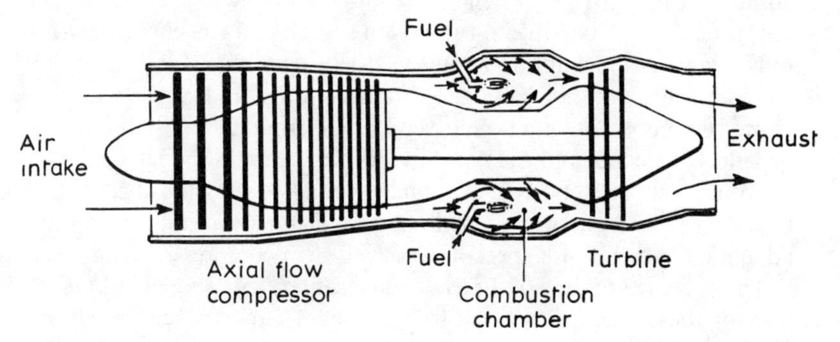

Fig. 11.7 Sketch of a typical turbo-jet engine fitted with an axial-flow compressor.

power outputs, corresponding to different rates of fuel flow, and must not be too sensitive to changes in the rate of flow, i.e. throttling.

The problem of the flammability limit is overcome by restricting the combustion to a small section of the air flow, the *primary air*, the burnt gases then mixing with cold *secondary air* to give the desired temperature. The fuel is atomized on entry to give a fine spray and reversal of flow is created so that burnt gas is recirculated through the flame to give continuous ignition and hence maintain stability. The simplest method is to direct the fuel into the approaching air. This is ineffective in high-velocity flows and it is more common to direct the fuel in the same direction as the overall flow but to contain it in a small chamber designed in such a way that the air flow is reversed inside. This chamber is cooled by the surrounding secondary air. The actual combustion occurs in a region of vortex flow, a typical burner being illustrated in Fig. 11.8.

In engines with centrifugal air compressors, a number of combustion chambers of the type illustrated in Fig. 11.8 are placed around the engine and the compressed air stream is directed by ducts to the individual chambers. In each chamber the inner flame tube is surrounded by a flow of secondary air which enters through slots and orifices in the flame tube walls. These openings are carefully designed to produce an air flow with the right degree of turbulence and recirculation. In a typical case the primary combustion air comprises about 30% of the total air flow, while the secondary dilution air makes up the remaining 70%.

Some axial compressor engines employ an annular combustion chamber which is open at one end to the compressor and at the other to

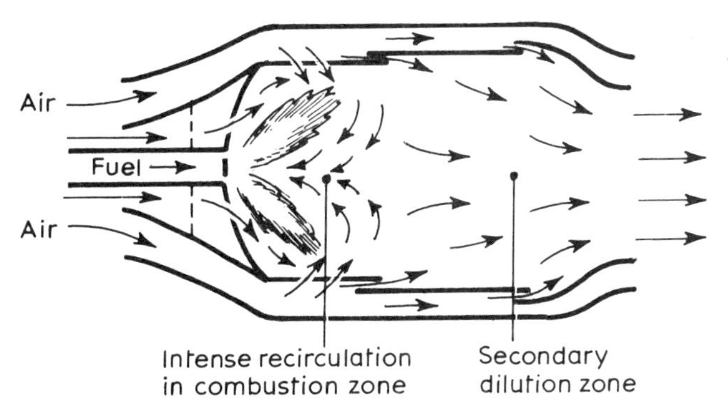

Fig. 11.8 Typical combustion chamber in a jet engine. The air flow through the chamber is indicated by arrows.

the turbine. Secondary cooling air flows around the inside and outside of the flame tube and holes in the flame tube walls allow cooling air to enter the tube.

11.4 SUGGESTIONS FOR FURTHER READING

Brame, J.S.S. and King, J.G. (1967) *Fuel, Solid, Liquid and Gaseous*, 6th edn, Edward Arnold, London.

Diesel Engines, Fuels and Lubricants, Esso Petroleum Company, London (1960).

Judge, A.W. (1967) *High Speed Diesel Engines*, 6th edn, Chapman and Hall, London.

Lewis, B. and Von Elbe, G. (1961) *Combustion, Flames and Explosions of Gases*, 2nd edn, Academic Press, New York.

Ricardo, H.R. and Hempson, J.G.G. (1968) *The High Speed Internal Combustion Engine*, 5th edn, Blackie and Sons Ltd, Glasgow.

Smith, M.L. and Stinson, K.W. (1952) *Fuels and Combustion*, McGraw-Hill, New York.

12

Heating applications

12.1 INTRODUCTION

A major use of combustion processes is in heating appliances, both industrial and domestic. The operation of such devices may be classified in terms of the burner used, which in turn depends on the physical state of the fuel employed:

(a) Gas burners: gaseous fuels are burnt in pre-mixed and diffusion flames.
(b) Liquid fuel burners: oil, the principal liquid fuel, is first atomized and the droplets burnt in an atmosphere of air.
(c) Dust burners: pulverized coal is dispersed in a gas stream and burnt in a similar manner to liquid fuels although, since the mechanism of combustion differs, the design of the combustion chamber is not identical.
(d) Fluidized-bed combustors: air is passed upwards through a bed of sand and coal particles which becomes fluidized. When the coal is ignited by some external means, a self-sustaining reaction takes place.
(e) Fuel beds: larger particles of solids fuels are also burnt simply by placing them on a grid through which air passes from below.

12.2 GAS BURNERS

The main aspects of gas burners have already been described in Chapter 3. All domestic and most industrial burners are simple variants of the Bunsen burner and provide an aerated flame. The gas enters a narrow tube through a jet and draws into it sufficient primary air to render the flame non-luminous. The remaining secondary air which is required to complete the combustion is entrained above the burner. In large industrial burners, it may be necessary to boost the

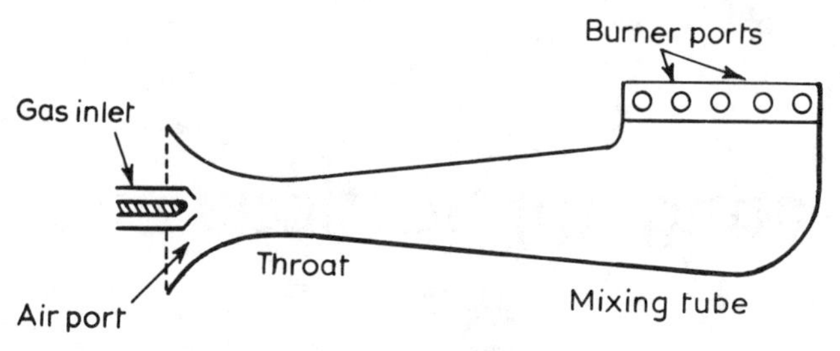

Fig. 12.1 Typical domestic gas burner.

pressure of gas above that of the mains and also to introduce the air under pressure.

A typical burner is illustrated in Fig. 12.1. The shape and diameter of the jet, the size of the air and burner ports, and the width of the throat all depend on the properties of the gas in use. Some of the factors affecting the design of gas burners have been mentioned briefly in Chapter 3.

12.3 OIL BURNERS

When oil is used for heating, it may be vaporized prior to burning, for example, by heating in a tube which passes through the flame, by spraying on to a hot surface, or by evaporation from a wick. Vaporizing burners of this type are used in small domestic appliances but are rarely used industrially because the cheap, heavy fuel oils tend to decompose before vaporizing.

In the alternative type of burner, the oil is 'atomized', or dispersed as fine droplets, in a stream of air. The droplet size usually covers a wide range, but with the majority of droplets less than 50 μm in diameter. Since the burning time for an oil droplet depends on its diameter (see Chapter 8), the size is kept small in order to complete combustion rapidly in a sufficiently small chamber volume. In addition, with larger droplets of the heavier oils, there is a tendency to form a solid carbon residue which can take many times longer than the liquid fuel to burn.

Atomization of the fuel oil may be achieved in two ways. In the *mechanical*, or *oil-pressure, atomizing burner*, the oil is forced at high pressure into the 'whirl' chamber through a series of tangential slots and then passes through an orifice into the combustion chamber. The static pressure is converted to rotational motion in the whirl chamber so that when the oil leaves the orifice the centrifugal force throws the oil outwards in a fine spray. In an *air atomizing burner*, streams of both air and fuel are brought together, the necessary dispersion being produced on mixing. The two streams may either pass through separate orifices and converge in the combustion chamber or they may be mixed first and then flowed through a single orifice.

12.4 PULVERIZED COAL BURNERS

Much of the world's electricity is generated by the combustion of pulverized fuel (p.f.). The coal is ground to particles about 40 μm in diameter and blown by a stream of air into the hot combustion chamber. This is a very large box lined with steel tubes containing water. The tubes are heated principally by radiation from the burning particles and the water in the tubes is converted to steam.

The burning mechanism of solid particles, including coal, has already been discussed in Chapter 8. In the first stage, the volatile material burns in a spherical diffusion flame surrounding the particle and, in the second, the solid residue undergoes burning at the surface. The combustion of the char is controlled by diffusion of oxygen in the porous structure, that is, Zone II kinetics apply, and only with large particles at very high temperatures does external mass transfer have any influence [170]. As a result, for the full range of fuels used for electricity generation, the times for burn-out (i.e. complete combustion) of typical particles do not differ by more than a factor of ten, whereas if Zone I kinetics were obeyed, the burn-out times would vary by a factor of up to one thousand. Because burn-out times cover such a relatively small range under the practical conditions of pulverized fuel combustion, basically the same technology can be used for all types of coal.

The flame speed in the dust cloud depends primarily on the combustion of the volatile portion of the material. The result is that flame speeds reach a maximum for fuel: air ratios 3 to 6 times the stoichiometric value (Fig. 12.2) [198]. In order to carry dust particles in a horizontal pipe, the air velocity must exceed 20 m s^{-1} and as the cloud passes through the nozzle into the combustion chamber it will expand and its speed will fall. As flame velocities of dusts are always below 20 m s^{-1}, the flame will stabilize itself somewhere ahead of the nozzle, In order to reduce the size of the combustion chamber and to operate under the most stable conditions, the fuel: air ratio is selected which gives the maximum flame velocity. Because this means using a very high ratio compared with the stoichiometric value, two separate air streams are used, the primary stream carrying the required amount of fuel and the secondary being used to complete the combustion. The

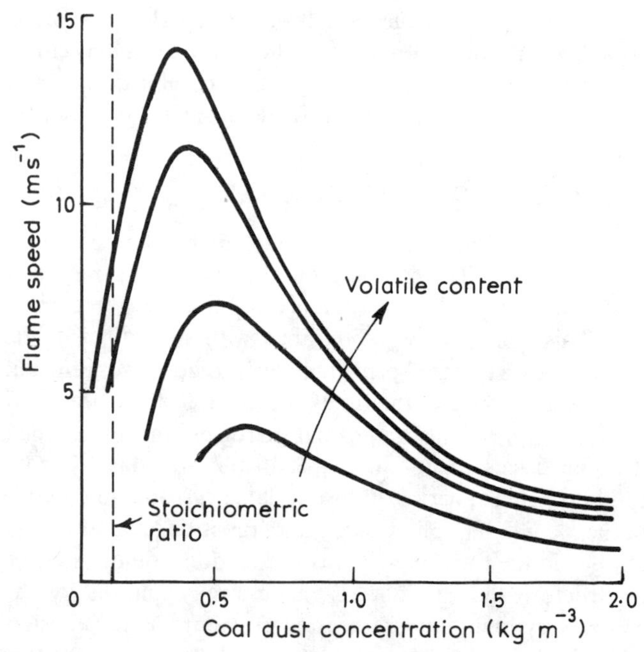

Fig. 12.2 Dependence of flame speed on coal dust concentration (after de Gray [198]).

way in which the primary and secondary air streams are mixed is, of course, an important factor in burner design. The optimum position to introduce the secondary air is probably that at which the first stage, the combustion of volatiles, is complete.

12.4.1 Cyclone furnace

Pulverized fuel burners suffer from certain disadvantages, the most important being that in order to obtain high heat release rates the fuel–air mixtures have to be passed in rapidly so that, for complete burning, large combustion chambers are required. In addition, the relative velocity of air to dust is small so that maximum burning rates are obtained only when a large amount of excess air is added. The ash tends to remain in the gas stream and can cause choking of exhaust flues.

Although the technological aspects of combustion have been largely omitted from this volume, it is worth referring briefly to the *cyclone furnace*. In this furnace, the combustible mixture is injected tangentially at high velocity (100–150 $m\,s^{-1}$). The burning particles are thrown outwards by centrifugal force and adhere to the molten slag on the walls. Although the throughput of air is very high, the residence time of the solid is extended so that combustion is completed. The slag can be tapped off from the walls and the relative motion between gas and solid ensures efficient mixing.

12.5 FLUIDIZED-BED COMBUSTION

A gas passed slowly upwards through a bed of solid particles finds its way through the spaces between the particles and the pressure drop across the bed is directly proportional to the flow rate. If the flow rate is increased, a point is eventually reached at which the frictional drag on the particles becomes equal to their apparent weight (i.e. weight minus any buoyancy force); the bed then expands as the particles adjust their positions to offer less resistance to the flow. The bed is now said to be *fluidized* and further increase in flow is not accompanied by an increase in pressure drop (Fig. 12.3). Above the minimum fluidization velocity, the extra gas over and above that required for fluidization passes through the bed as bubbles and the bed itself is agitated.

In a fluidized bed, heat and mass transfer between the fluidizing gas and the solid particles are extremely efficient and fluidized-bed reactors are used for carrying out many chemical reactions on an industrial scale (e.g. catalytic cracking of hydrocarbons) [199]. Combustion in a fluidized bed has some particular attractions, including its suitability for use with low-grade, high-ash coals and the lower bed temperatures compared with those in a conventional furnace. As a result of this low

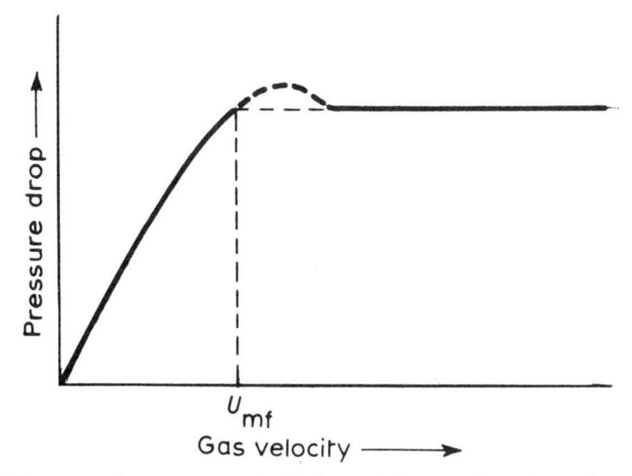

Fig. 12.3 Pressure drop across a bed of particles as a function of gas velocity. u_{mf} corresponds to the minimum fluidization velocity.

temperature the NO_x levels in the flue gas are considerably reduced. A marked reduction in sulphur dioxide emissions can also be achieved by mixing the coal with limestone (or dolomite); at the temperature of the bed this is converted to the oxide with which the sulphur dioxide reacts to give calcium (or magnesium) sulphate

$$CaO + SO_2 + \tfrac{1}{2}O_2 \rightarrow CaSO_4 \tag{12.1}$$

The solid sulphate is discharged along with the normal ash. An additional advantage of fluidized-bed combustors is that they can use fairly coarse coal particles (~ 1 mm diameter) and there is no need for much of the costly crushing equipment associated with the preparation of pulverized fuel.

Unfortunately, the fluid mechanics of fluidized beds are, as yet, incompletely understood and so the scale-up of laboratory units to industrial size is very difficult. A fluidized bed resembles a vigorously boiling liquid and the most widely used theory envisages two phases, a *lean* (or *bubble*) phase and a *dense* (or *emulsion* or *particulate*) phase which closely resembles the bed at the state of incipient fluidization [200]. The fluidizing gas passes through the bed in the dense phase and the excess in the lean phase. Gas is exchanged between these two phases and the exact nature of the flow patterns in and around the rising bubbles has been the subject of much investigation. In a fluidized-bed combustor used for steam-raising, water tubes are immersed in the bed

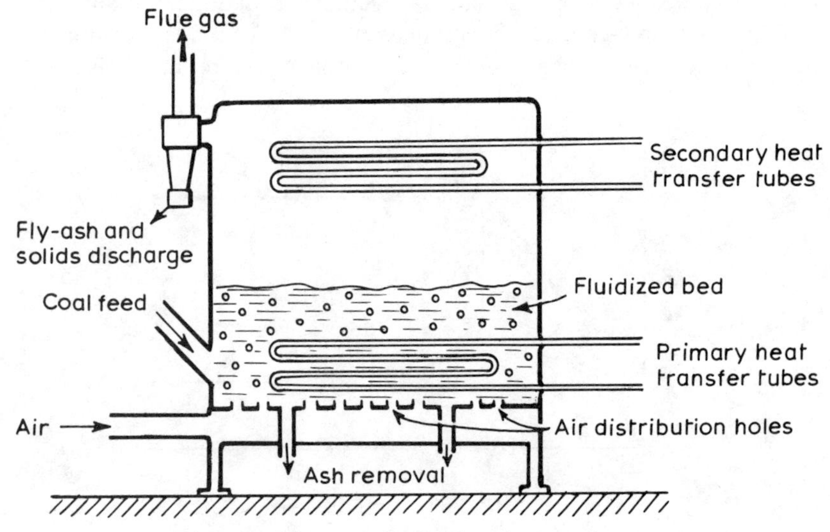

Fig. 12.4 Diagram of a fluidized-bed combustor.

and the flow of bubbles through and around banks of these tubes is a further complication in the design of the unit.

Fig. 12.4 shows the essential features of a typical medium-sized fluidized bed combustor. Normally the bed consists initially of sand or some similar inert material which is fluidized by an air stream and raised to the ignition temperature by some external heating source. When the requisite temperature is reached, coal is fed to the vigorously bubbling bed where it becomes thoroughly mixed with the sand. As combustion begins, volatile material is given off and usually burns in the freeboard above the bed. The solid residue, or char, remains in the bed where the temperature is about 1175 K. The calculation of the burn-out time for various types of coal particle requires knowledge of both the chemical kinetics and the processes whereby gas is transferred from the bubbles to the dense phase to complete the combustion of the char. At present our understanding of the latter is imperfect and this is one of the factors which makes the design of large-scale units so difficult.

Ash from the coal and the residue from the limestone added to remove sulphur, together with the initial sand, are removed from the bottom of the bed, although char particles eventually become small enough to be carried over in the flue gases. This *fly ash* has to be removed before these gases are discharged to the atmosphere. Since the unburned material represents a significant loss of combustion efficiency, arrangements are often made to recirculate it.

Fluidized-bed combustors are capable of very high rates of heat release ($\sim 1\ \mathrm{MW\,m^{-3}}$) but these demand high fluidizing velocities and large particles compared with those in fluidized bed reactors used to carry out normal chemical processes. The bubbles are relatively small and recirculation of gas within them, which is an important feature of other fluidized bed reactors, is not so pronounced.

12.6 SOLID FUEL BEDS

The best-known example of coal combustion is provided by the domestic fire and the majority of industrial coal burners operate on the same overfeed principle. Once combustion has commenced, a number of separate reaction zones can be distinguished (Fig. 12.5). The fresh, or 'green', coal is placed on the upper surface and the heat transferred from the bed below causes evaporation of the volatile material which burns in the secondary air and leaves a residue of fixed carbon or coke.

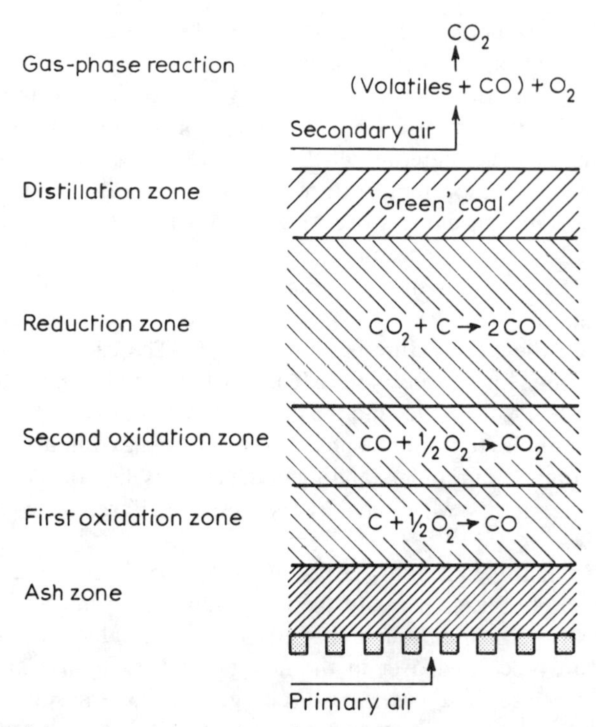

Fig. 12.5 Schematic representation of combustion processes occurring in a solid fuel bed.

Primary air enters the base of the grate and passes first through the ash zone. The ash performs a useful function in providing an insulation between the high-temperature reaction zone and the grate. In the first oxidation zone, the oxygen reacts at the surface of the carbon to give carbon monoxide

$$C + \tfrac{1}{2}O_2 \rightarrow CO \qquad (12.2)$$

Carbon monoxide is released from the solid and reacts on mixing with oxygen, in the second oxidation zone, to give carbon dioxide

$$CO + \tfrac{1}{2}O_2 \rightarrow CO_2 \qquad (12.3)$$

Each of these reactions is strongly exothermic and some of the heat released is transferred back to promote the reaction. This is usually the hottest part of the fuel bed.

At this stage the oxygen concentration is very much depleted and the carbon dioxide is reduced by the Boudouard reaction in the next layer

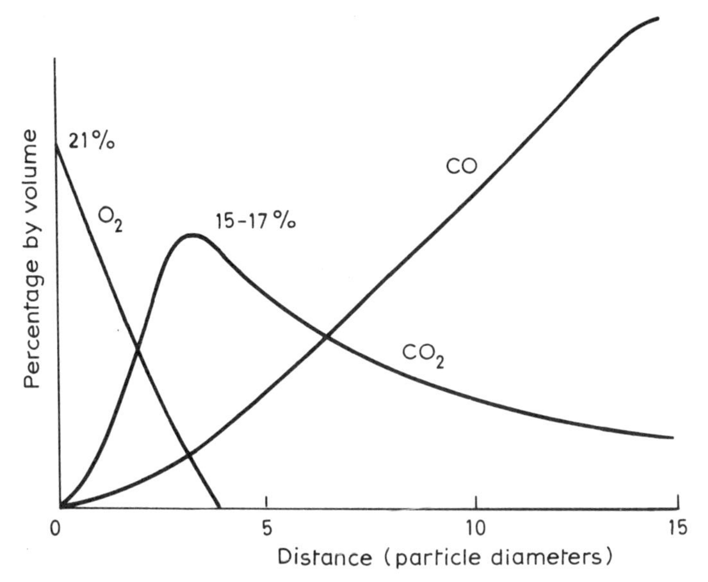

Fig. 12.6 Variation of gas composition through a solid fuel bed (after Thring [201]).

of fuel

$$CO_2 + C \rightarrow 2CO \quad (12.4)$$

As this reaction is endothermic, the temperature will begin to fall. When the carbon monoxide leaves the fuel, it mixes with secondary air and burns again to carbon dioxide. The concentration changes through the zones are illustrated in Fig. 12.6 [201]. It will be seen that the oxidation zone is quite small, corresponding to only a few particle diameters, whilst the reduction zone is thicker because the reaction rate there is much slower. The maximum temperature coincides roughly with the position of the maximum in the CO_2 concentration.

The relative importance of the various factors which control the rate of combustion depend on the temperature of the bed and, in principle, Zone I, II or III kinetics as defined in Chapter 8 may apply [170].

These considerations apply both to the primary reaction of carbon with oxygen and to the Boudouard (CO_2 reduction) reaction. It appears that, at the temperatures existing in fuel beds, the control is exercized mainly by chemical reaction rather than by diffusion.

In the high-temperature region (Zone III) the reaction rate will depend very much on the aerodynamics of the flow. Clearly, the boundary layer thickness cannot exceed the radius of the channels between the particles. The thickness of this diffusion layer will also be reduced considerably if the flow is turbulent and a transition in burning rates has been observed which corresponds to such a change. In a practical fuel bed, it is desirable to have well-graded fuel otherwise the variation in channel size leads to uneven reaction and reduced efficiency.

A major aim of this text has been to demonstrate how the phenomena of combustion depend on a complex interplay of physical and chemical processes. It is perhaps fitting to conclude this chapter by taking as an example the ignition and extinction characteristics of the domestic coal fire.

It is common experience that an ignition source of a certain intensity is required for burning to be sustained and alternative efforts prove of little avail in continuing the combustion. Once a certain temperature is attained burning continues and is then very susceptible to forced draught, the intensity increasing dramatically with increasing air velocity. However, beyond a certain critical air flow, the combustion is sharply extinguished and does not necessarily recommence at the same point when the air flow is reduced.

These phenomena may be explained [170,202] in terms of a very simple model. When true combustion occurs, the system is self-heating in a manner directly analogous to a thermal explosion and the rate is controlled primarily by the rate of diffusion of oxygen to the surface. This is then a region of diffusion control and the rate varies directly with the rate of flow of air.

Extinction occurs, and combustion ceases, when the heat losses become so great relative to the heat production that the chemical reaction rate falls below the diffusion rate and chemical or kinetic control takes over. At the high-velocity limit, the heat loss is due to convective transfer, the energy being carried away by the products, and at the lower limit, the loss is due to radiant transfer.

It should encourage all who seek to apply fundamental scientific principles to practical problems that this apparently complicated state of affairs can be understood quantitatively with the aid of the relatively simple ideas outlined in the preceding chapters.

12.7 SUGGESTIONS FOR FURTHER READING

Brame, J.S.S. and King, J.G. (1967) *Fuel, Solid, Liquid and Gaseous*, 6th edn. Edward Arnold, London.

Howard, J.R. (ed.) (1983) *Fluidized Beds: Combustion and Applications*, Applied Science Publishers, Barking.

Mulcahy, M.F.R. (1978) The combustion of carbon in oxygen, in *The Metal and Gaseous Fuel Industries*, The Chemical Society, London.

Smith, M.L. and Stinson, K.W. (1952). *Fuels and Combustion*, McGraw-Hill, New York.

Thring, M.W. (1962) *The Science of Flames and Furnaces*, 2nd edn, Chapman and Hall, London.

Vulis, L.A. (1961) *Thermal Regimes of Combustion* (translated by M.D. Friedman), McGraw-Hill, New York.

13 Combustion and the environment

13.1 COMBUSTION-GENERATED POLLUTION

The products of complete combustion of hydrocarbon fuels are carbon dioxide and water. However, in the exhaust gases from any practical device there will be in addition to these compounds, carbon monoxide, nitrogen oxides ('NO_x'), unburned hydrocarbons and probably solid particulates (soot). Depending on the circumstances other materials may be present, for example, lead compounds from antiknocks added to motor fuel, or sulphur dioxide from the sulphur present in coal or heavy fuel oil. The 'lifetime' of these species in the atmosphere is relatively short and if they were distributed evenly their harmful effects would be minimal. Unfortunately these man-made effluents are usually concentrated in localized areas and their dispersion is limited by meteorological and topographical factors. Furthermore, synergistic effects mean that the pollutants interact with each other: in the presence of sunlight, carbon monoxide, NO_x and unburned hydrocarbons lead to *photochemical smog*, while when sulphur dioxide concentrations become appreciable, *SO_2-based smog* is formed. The first of these is typical of a Los Angeles smog and the second was associated with London in the era prior to the *Clean Air Acts* (1956 and 1968) when inefficient coal burning appliances were widely used.

Smog formation is encouraged by the geography of the Los Angeles basin; the conurbation is encircled by mountains and the sea, a situation which is favourable to temperature inversions which trap the heavily polluted, cool surface air below a layer of warmer air. The intensity of a photochemical smog is at its worst near midday when the temperature is high (*ca.* 25° C), the sky clear and the relative humidity low. The main effects are eye irritation, reduced visibility and damage to plant life. In contrast, London smogs occurred at times when the temperature was close to 0° C and the relative humidity was high; the most noticeable immediate effect on man was the fog and its accompanying severe bronchial irritation, but there was also long-term damage to buildings and other property.

It is worth remembering that even the non-toxic carbon dioxide may have an important effect on the environment. The Earth's surface emits infrared radiation with a peak of energy distribution in the region 13–18 μm where carbon dioxide is a strong absorber. This results in the *greenhouse effect* whereby this infrared radiation is trapped by the atmosphere and the temperature of the Earth's surface is substantially raised. As a result of the combustion of fossil fuels, the concentration of carbon dioxide in the atmosphere is increasing from its present level of about 300 ppm at the rate of about 1 ppm per year. Although many factors are involved, it does seem that a doubling of the carbon dioxide concentration in the atmosphere would result in an average temperature change at the surface of about 2 K. A temperature increase of this magnitude would bring about an appreciable reduction in the polar ice-caps and this in turn would result in further heating. While it is true that the present rate of increase of carbon dioxide is fairly low, nevertheless, if by the end of the twenty-second century most of the world's known fossil fuel supplies are consumed, it is estimated that the amount of carbon dioxide in the atmosphere will have risen by a factor of between 4 and 8 from its present level. A change of this magnitude must bring about substantial alterations in the Earth's climate [203].

13.1.1 Formation of pollutants in internal combustion engines

(a) *Carbon monoxide*

The current generation of spark-ignition internal combustion engines operate with slightly fuel-rich mixtures (equivalence ratio, Φ, between 1.1 and 1.4) since this gives smoother running, good power characteristics and avoids knocking. Under normal running conditions, the exhaust gases from such an engine contain between 1 and 2% carbon monoxide but this level may rise under idling or decelerating conditions. These carbon monoxide concentrations can be considerably reduced if the mixture is made leaner (Fig. 13.1). Thus, at $\Phi \approx 0.95$, the carbon monoxide level may be down to 0.5%. Conventional spark-ignition engines do not run satisfactorily with such lean mixtures and considerable attention has been directed towards the development of new types of engine which utilize fuel-lean mixtures; some of these have been mentioned in Chapter 11.

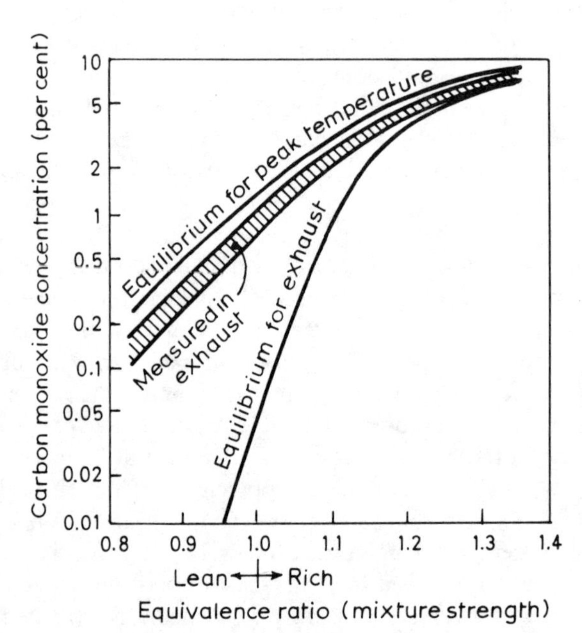

Fig. 13.1 Carbon monoxide concentration in engine exhaust. Measurement and equilibrium prediction (after Starkman [204]).

Under the conditions obtaining at the peak of the engine cycle (*ca.* 2800 K and 30 atm) the equilibrium

$$CO + OH \rightleftharpoons CO_2 + H \quad (13.1)$$

is quickly established. As the temperature falls the position of equilibrium moves to the right and the concentration of carbon monoxide in the exhaust gas should be quite small but the measured levels are closer to those at peak cycle conditions (Fig. 13.1). The reason why the equilibrium is 'frozen' at its high temperature position is not properly understood. The rate of the forward reaction is very fast and does not change appreciably with temperature [205], while the supply of OH radicals is maintained by the reactions

$$H + O_2 \rightleftharpoons OH + O \quad (13.2)$$

$$O + H_2 \rightleftharpoons OH + H \quad (13.3)$$

$$OH + H_2 \rightleftharpoons H_2O + H \quad (13.4)$$

and may even exceed the equilibrium value because during the expansion part of the cycle third-order recombinations such as

$$H + OH + M \rightarrow H_2O + M \tag{13.5}$$

are greatly slowed by the fall in pressure, relative to bimolecular processes. Possible explanations [196] are that carbon monoxide formed when the flame is quenched near the cylinder walls is a significant contribution to the total carbon monoxide and that partial oxidation of unburned hydrocarbons during the exhaust part of the cycle increases the measured carbon monoxide concentrations in the effluent.

(b) *Nitrogen oxides*

In the exhaust gases NO_x is almost entirely nitric oxide, NO. A typical level in exhaust gas is 1000 ppm (0.1 mol%) and this exceeds the concentration corresponding to equilibrium at the maximum temperature in the cylinder (Fig. 13.2). The mechanism of nitric oxide formation is based on the Zeldovich scheme. In the combustion chamber, the

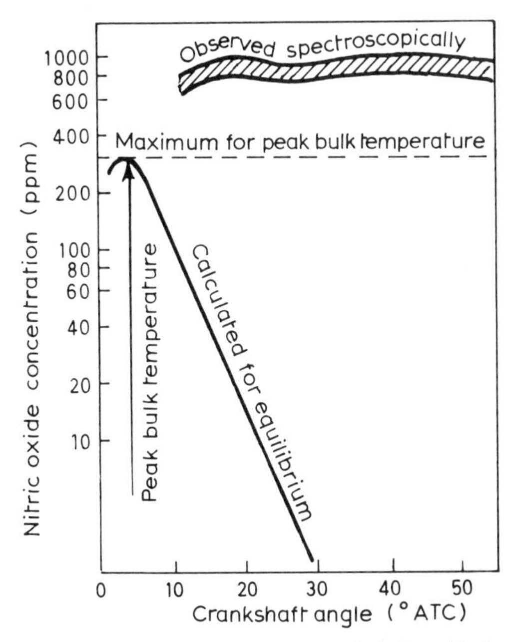

Fig. 13.2 Optical spectrometer determination of nitric oxide in an operating engine. Equivalence ratio = 1.16, speed = 1294 rpm. Ignition 38° BTDC (after Starkman [204]).

equilibrium

$$O_2 \rightleftharpoons 2O \tag{13.6}$$

is established and this is followed by the linear chain steps

$$O + N_2 \rightleftharpoons NO + N \tag{13.7}$$

$$N + O_2 \rightleftharpoons NO + O \tag{13.8}$$

There is also a contribution from

$$N + OH \rightleftharpoons NO + H \tag{13.9}$$

Thus, nitric oxide is formed mainly in the post-flame gases where the O atom concentration and the temperature are high. There are appreciable concentration gradients across the combustion chamber and nitric oxide concentrations are especially high around the sparking plug where the temperature remains high for the longest period.

Kinetic factors, rather than equilibrium considerations, control nitric oxide levels in the exhaust gas. Reaction 13.7 has a high activation energy ($303\,kJ\,mol^{-1}$) in the forward direction and consequently any reduction in the flame temperature sharply reduces the rate of nitric oxide formation. For this reason, injection of water into the combustion chamber of both spark-ignition and diesel engines has been suggested as a means of reducing nitric oxide levels in exhaust gases. Very considerable reductions in nitric oxide emissions have been achieved in this way [193,194], but at present the economic penalties accompanying the additional complications have prevented its wide adoption on a commercial scale.

The nitric oxide concentration in the exhaust gases depends on the mixture composition and reaches a peak on the lean side of stoichiometric. Unfortunately, this conflicts with the optimum conditions for carbon monoxide formation (Fig. 13.3).

There is growing evidence for another path to nitric oxide formation, the so-called *prompt-NO* mechanism [206]. Although the matter is far from being fully resolved at the present time the mechanism probably involves HCN which has been detected in hydrocarbon flames. This compound may be formed by either

$$CH + N_2 \rightarrow HCN + N \tag{13.10}$$

or

$$CH_2 + N_2 \rightarrow HCN + NH \tag{13.11}$$

Distinguishing between these alternatives is difficult because in the

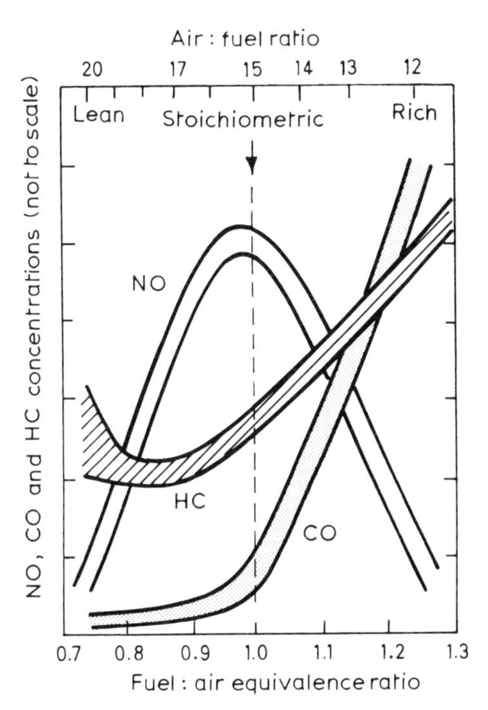

Fig. 13.3 Variation of HC, CO and NO concentration in the exhaust of a conventional SI engine with fuel–air equivalence ratio. HC = unburned hydrocarbon (after Heywood [196]).

flame the N and NH concentrations will be balanced by the equilibrium

$$NH + OH \rightleftharpoons N + H_2O \tag{13.12}$$

and a further possibility is that both Reactions 13.10 and 13.11 occur. CN species do not form NO directly. The first step is probably

$$HCN + OH \rightarrow HOCN + H \tag{13.13}$$

followed by the conversion of HOCN to HNCO and eventually species of the type NH_i ($i = 0$–3) which are oxidized to NO [207]. If this pathway is substantiated, it may account for the 'excess' nitric oxide which cannot be explained by the Zeldovich mechanism.

(c) *Unburned hydrocarbons*

Exhaust gases are commonly said to contain unburned hydrocarbons. In fact, with spark-ignition engines there is little correspondence between the chemical composition of the unburned material in the exhaust and the original fuel. Not only smaller alkanes and alkenes, but also carbonyl compounds and peroxides are present in engine exhaust and the term 'unburned hydrocarbons' usually implies all the not completely reacted fuel. Many of these compounds undergo rapid reaction in the atmosphere and are ultimately involved in the formation of photochemical smog (see below). Important sources of unburned hydrocarbons are the *quench layer* and the *crevices* between the piston and the cylinder wall (Fig. 13.4). As the flame propagates across the combustion chamber and approaches the wall, heat transfer and radical destruction processes quench the flame close to the wall. Although the flame has not traversed this quench layer, appreciable reaction takes place resulting in the formation of products by the partial oxidation and pyrolysis of the fuel: the thickness of the quench layer depends on many factors but is typically around 0.1 mm.

Recent work [208] has emphasized that the quench layer is by no means the only source of unburned hydrocarbons in the exhaust gas. Between one-quarter and one-half of the total emissions of unburned hydrocarbons comes from the crevices around the piston rings. Lubricating oil on the cylinder walls is the origin of a similar amount, the remainder coming from oil stored in carbon deposits on the piston crown and elsewhere, from miscellaneous crevices and leakage past the exhaust valve, as well as from the quench layer itself.

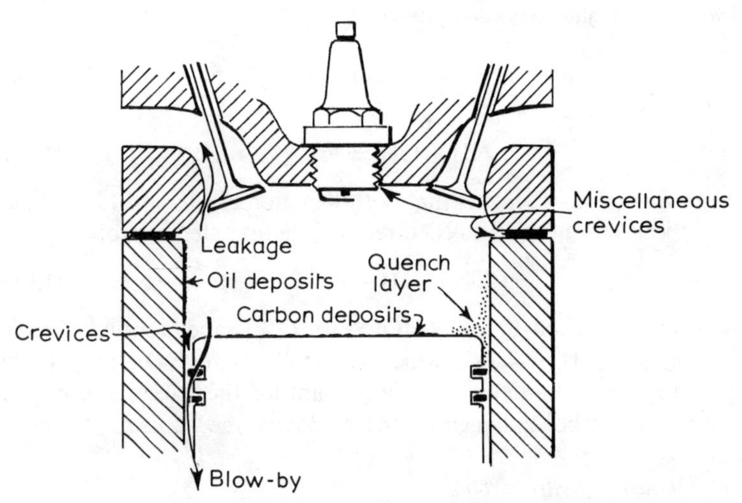

Fig. 13.4 Diagram showing the principal sources of unburned hydrocarbons in engine exhaust emissions.

Unburned hydrocarbon concentrations in the exhaust gas decrease as the mixture is made leaner, but eventually, with very lean mixtures, misfiring results in a rapid increase in the unburned material in the exhaust (Fig. 13.3).

In a diesel engine there is no moving flame front as there is in a spark-ignition engine. The fuel–air mixture burns wherever the conditions of composition and temperature are suitable. Although the fuel is injected as a spray, the evaporation rate is normally fast enough to ensure homogeneous rather than heterogeneous droplet combustion except where the spray strikes the walls. Most diesel engines, if correctly operated and maintained, emit less carbon monoxide and unburned hydrocarbons than comparable spark-ignition engines; they tend, however, to emit more smoke, particulates and compounds such as aldehydes with objectionable odours [210].

(d) *Smoke and particulates*

The visible material in exhaust gases (smoke) includes carbon particles and droplets of unburned hydrocarbon. Polynuclear aromatic hydrocarbons are also present and these are highly undesirable since many of them are carcinogenic.

The mechanism of soot formation has been discussed in Chapter 7. Although large soot particles are fairly resistant to oxidation, small ones are removed by reaction with hydroxyl radicals

$$C_n + OH \rightarrow CO + C_{n-1} + H \qquad (13.14)$$

This reaction is fast and the concentration of OH cannot be maintained since the only source in the exhaust gases is the dissociation reaction (13, −5) which has a large activation energy. If an alkaline earth metal is added, the supply of OH is facilitated by, for example,

$$Ba(OH)_2 + M \rightarrow BaO + H_2O + M \qquad (13.15)$$

$$BaO + H_2 \rightarrow BaOH + H \qquad (13.16)$$

$$BaOH + H_2O \rightarrow Ba(OH)_2 + H \qquad (13.17)$$

$$H + H_2O \rightarrow OH + H_2 \qquad (13.18)$$

and such additives are very effective in reducing smoke emission even under fuel-rich conditions. Other metals are effective in lean mixtures and probably function by catalysing various heterogeneous processes.

In the spark-ignition engine, lead antiknocks are converted to solid lead oxide (PbO) and particles of this compound, together with sulphides and halides of lead, are detectable in the exhaust gas. The nature of the lead-containing particles depends on several factors: in city streets there is already a background aerosol and the lead is found as coagulated, chain-aggregate particles, whereas from engines running at high speed as, for example, on a motorway, the lead is emitted as very small discrete particles about 0.02 μm in diameter.

The lead levels in urban atmospheres may be about 3 $\mu g\,m^{-3}$, compared with a background of about 0.8 $\mu g\,m^{-3}$; close to heavily used motorways atmospheric lead may be up to 15 $\mu g\,m^{-3}$. About one-tenth of the total lead emitted is deposited fairly close to the roadside, the remainder being dispersed widely over the countryside or even further afield [211].

13.1.2 Suppression of pollutants from internal combustion engines

The requirements for reducing undesirable emissions from engines are conflicting, not only amongst themselves in some cases (Fig. 13.3), but also with the need to ensure 'drivability' of the vehicle and reasonable fuel economy and performance.

A great deal can be done to minimize emissions by good maintenance and engine design and it has been shown that with conventional carburettors poor mixture distribution between the cylinders can lead to partial misfiring and even condensation in the inlet manifold. When the throttle is suddenly closed as in deceleration, most of this fuel is rapidly vaporized and passes as an over-rich mixture through the engine thus contributing to the exhaust emissions. By eliminating variations in the air:fuel ratio between cylinders, engines can run on much leaner overall mixtures with marked reductions in the CO and NO_x emissions, although unburned hydrocarbons are relatively unaffected.

Engines using stratified charge combustion chambers or pre-chambers have the advantage of being able to run on lean mixtures which results in lower carbon monoxide and unburned hydrocarbon concentrations in the exhaust gases. Even so, some emissions remain and attention has been directed to ways of reducing these still further. One of these which successfully reduces nitric oxide is exhaust-gas recycle (EGR). This involves dilution of the fuel–air mixture by adding up to 20% exhaust gas. The result is lower combustion temperatures and a reduction of nitric oxide to less than half its normal amount, though carbon monoxide and unburned hydrocarbon levels are not improved and may even be adversely affected.

At present perhaps the most widely used device fitted to production vehicles is the catalytic converter to control carbon monoxide and unburned hydrocarbons. The exhaust gases flow through the catalyst bed before emerging to the atmosphere at about 250° C. As emissions are highest when the engine is cold and the choke is used for starting, it is necessary to ensure that the catalyst heats up as quickly as possible. Noble metals, for example, platinum, palladium or rhodium, on an inert support are used as catalysts and as these are poisoned by lead this is an additional reason why lead compounds are being replaced by organic antiknocks such as methyl *t*-butyl ether.

Nitric oxide can also be removed catalytically by reduction to nitrogen; in this case, the engine must be operated with a fuel-rich mixture and the nitric oxide removed first from the virtually oxygen-free exhaust gases before additional air is added and the exhaust gases pass through a second catalytic bed where carbon monoxide and hydrocarbons are burned.

13.1.3 Pollutants from power stations and other industrial sources

Although gasoline and diesel fuels have a low sulphur content (~ 0.2%), a typical coal contains 2% sulphur while a heavy fuel oil may have double this amount. Very large amounts of coal and fuel oil are consumed in generating electricity and the sulphur emissions from power stations and similar industrial sources represents a considerable environmental hazard, especially as much of the sulphur is returned to the Earth's surface as acid rain, often many hundreds of kilometres from its original source [212]. The sulphur in the fuel is converted to sulphur dioxide

$$S + O_2 \rightarrow SO_2 \qquad (13.19)$$

and sulphur dioxide levels in the flue gas from large furnaces and boilers may be as high as 2000 ppm. At flame temperatures in the presence of excess air, some sulphur trioxide is also formed

$$SO_2 + O + M \rightarrow SO_3 + M \qquad (13.20)$$

Only a small amount of sulphur trioxide can have an undesirable effect as it brings about the condensation of sulphuric acid and causes severe corrosion. Although diminution of the excess air reduces sulphur trioxide formation considerably, other considerations, for example, soot formation, dictate that the excess air level cannot be lowered sufficiently to eliminate sulphur trioxide entirely.

While it is technically feasible to reduce the sulphur content of oils and even to some extent that of coal, it is not economic to attack the problem of sulphur dioxide and trioxide emission by this means alone. A more attractive proposition involves the introduction of finely ground dolomite or limestone which removes the acid gases. This is particularly easy in fluidized-bed combustion where the powdered limestone can be added to the fuel feed and the calcium sulphate removed continuously with the ash. Other methods involve scrubbing the gas with the slurry of a suitable basic material, adsorption on carbon or alumina, or catalytic conversion to hydrogen sulphide which is absorbed in ethanolamine solution.

The non-volatile oxidation products of the inorganic constituents of the fuel (e.g. Na, K, V compounds) are left as ash and help to form corrosive deposits in many cases. With atomized fuel and pulverized fuel combustion, a large proportion of the ash is carried out of the combustion chamber with the exhaust gas stream and this fly ash has to be removed as far as possible usually in electrostatic precipitators: even then, a small amount escapes to the atmosphere where the solid particles act as condensation nuclei and surfaces where sulphur dioxide is converted to the trioxide.

Good design and operation of large furnaces can reduce solid emissions and keep the carbon monoxide level in the flue gas below 0.01%. Nitric oxide is formed by the Zeldovich mechanism, from nitrogenous compounds in the fuel and probably also by the prompt-NO route. Many means have been tried to reduce nitric oxide emissions, including low excess air which was first applied in oil-fired power stations as a means of combating corrosion from sulphur oxides. In a boiler or furnace, the combustion products spend a relatively long time at flame temperatures and consequently NO formation by the Zeldovich route is favoured. Low excess air keeps the oxygen atom concentration down and minimizes Zeldovich-NO, provided that adequate heat transfer is maintained to prevent increased flame temperatures. Fluidized-bed combustion generates virtually no thermal-NO despite the long residence times since the bed temperatures are far too low and the residual NO is derived from the fuel nitrogen rather than via the prompt mechanism.

13.1.4 Reactions of pollutants in the atmosphere

Atmospheric chemistry is a vast and rapidly growing subject and in the space available the discussion must be limited to topics directly related to combustion. Mention must, however, be made of the fact that other man-made materials such as halogenated hydrocarbons from aerosols

Table 13.1 Typical concentrations of the major pollutants in photochemical smog [214].

Pollutant	Concentration range (pphm)*
Carbon monoxide	200–2000
Nitric oxide	1–15
Nitrogen dioxide	5–20
Total hydrocarbons (excluding methane)	20–50
Alkenes	2–6
Aromatics	10–30
Ozone	2–20
Aldehydes	5–25
Peroxyacyl nitrates	1–4

*pphm = parts per hundred million

and nitrogenous compounds from fertilizers are also present in the atmosphere and these also play a part in the complex chemistry which takes place there [213].

The primary pollutants emitted by internal combustion engines and industrial sources undergo reaction in the atmosphere to produce secondary pollutants which give rise to smog. The most important secondary pollutants are nitrogen dioxide (NO_2), ozone (O_3), aldehydes, ketones, peroxyacyl nitrates and alkyl nitrates, and their effects are eye irritation, damage to plant life and a reduction in visibility. Table 13.1 gives typical concentrations of these pollutants in a photochemical smog.

Nitrogen oxides play an important role in air pollution. The primary pollutant, nitric oxide, does not absorb sunlight and the gas-phase reaction

$$2NO + O_2 \rightarrow 2NO_2 \tag{13.21}$$

is third order and quite slow. The conversion to nitrogen dioxide is brought about in other ways. Nitrogen dioxide absorbs solar radiation ($\lambda \leqslant 398$ nm) producing ground-state oxygen atoms by photodissociation

$$NO_2 + h\nu \rightarrow NO + O(^3P) \tag{13.22}$$

This is followed by

$$O(^3P) + O_2 + M \rightarrow O_3 + M \tag{13.23}$$

$$O_3 + NO \rightarrow O_2 + NO_2 \quad (13.24)$$

and the net effect is the establishment of the equilibrium

$$NO_2 + O_2 \rightleftharpoons NO + O_3 \quad (13.25)$$

Although Reaction 13.23 is not fast, the high concentration of oxygen molecules in the atmosphere means that it is the main pathway by which oxygen atoms react. Some oxygen atoms and ozone molecules react with hydrocarbons, especially alkenes and alkyl-benzenes, present in a polluted atmosphere initiating a complex sequence of reactions leading to the formation of undesirable secondary pollutants. However, these reactions alone are not fast enough to account for the observed behaviour and it now seems clear that OH radicals are also involved. In a non-polluted atmosphere, the OH concentration is maintained by a chain involving electronically excited oxygen atoms

$$O_3 + h\nu \xrightarrow{\lambda < 310\,\text{nm}} O_2 + O(^1D) \quad (13.26)$$

$$O(^1D) + H_2O \rightarrow 2OH \quad (13.27)$$

Most excited oxygen atoms do not react in this way and are quenched to their ground state

$$O(^1D) + M \rightarrow O(^3P) + M \quad (13.28)$$

and then recombine with O_2 to give ozone by Reaction 13.23. In addition, HO_2 radicals are formed by photolysis of formaldehyde, traces of which derived from naturally formed methane are normally present in the atmosphere (see below).

$$HCHO + h\nu \xrightarrow{\lambda < 338\,\text{nm}} H + HCO \quad (13.29)$$

$$H + O_2 + M \rightarrow HO_2 + M \quad (13.30)$$

$$HCO + O_2 \rightarrow HO_2 + CO \quad (13.31)$$

There is a second channel in formaldehyde photolysis which leads to hydrogen formation

$$HCHO + h\nu \rightarrow H_2 + CO \quad (13.32)$$

This channel is strongly pressure- and wavelength-dependent.

Nitric oxide also reacts rapidly with HO_2 radicals

$$NO + HO_2 \rightarrow NO_2 + OH \quad (13.33)$$

leading to production of hydroxyl radicals and more ozone by Reactions 13.22 to 13.24, while ozone is destroyed by

$$O_3 + HO_2 \rightarrow 2O_2 + OH \tag{13.34}$$

as well as by photolysis as shown in Reaction 13.26.

The interconversion of OH and HO_2 occurs by fast bimolecular reactions with other trace components in the atmosphere, for example,

$$OH + CO \rightarrow H + CO_2 \tag{13.35}$$

$$OH + H_2 \rightarrow H + H_2O \tag{13.36}$$

followed by Reaction 13.30 in each case, as well as by Reaction 13.33. Carbon monoxide is not, therefore, an inert component in the atmosphere, but reacts rapidly by Reaction 13.35 which is one of the most important processes in lower atmospheric chemistry.

Methane is normally present in the atmosphere as a result of natural gas leakage, natural fermentation and other biological processes, as well as being produced by incomplete combustion. Hydroxyl radical attack on methane

$$OH + CH_4 \rightarrow H_2O + CH_3 \tag{13.37}$$

is rapidly followed by

$$CH_3 + O_2 + M \rightarrow CH_3O_2 + M \tag{13.38}$$

$$CH_3O_2 + NO \rightarrow CH_3O + NO_2 \tag{13.39}$$

$$CH_3O + O_2 \rightarrow HCHO + HO_2 \tag{13.40}$$

$$CH_3O_2 + CH_3O_2 \rightarrow CH_3OH + HCHO + O_2 \quad \text{or} \quad 2CH_3O + O_2 \tag{13.41}$$

When the nitric oxide concentration is very low other reactions of methylperoxy radicals become important, especially

$$CH_3O_2 + HO_2 \rightarrow CH_3OOH + O_2 \tag{13.42}$$

followed by

$$CH_3OOH + h\nu \rightarrow CH_3O + OH \tag{13.43}$$

The production of alkoxy and alkylperoxy radicals by means such as Reactions 13.39 and 13.38 opens up further pathways which can account for the formation of methyl nitrate and peroxyacetyl nitrate (PAN), both of which are found in irradiated polluted atmospheres. The former is produced by

$$CH_3O + NO_2 \rightarrow CH_3ONO_2 \tag{13.44}$$

while one route to PAN is via acetaldehyde which itself is formed by the photoxidation of hydrocarbons such as but-2-ene

$$CH_3CHO + OH \rightarrow CH_3CO + H_2O \tag{13.45}$$

$$CH_3CHO + h\nu \rightarrow CH_3 + HCO \tag{13.46}$$

$$CH_3CO + O_2 \rightarrow CH_3CO_3 \tag{13.47}$$

$$CH_3CO_3 + NO_2 \rightarrow CH_3CO{\cdot}OONO_2(PAN) \tag{13.48}$$

Thus, there are several complex interlocking chains propagated by the OH and HO_2 radicals and their organic analogues RO and RO_2 where R can be alkyl, acyl or any other carbon-containing fragment. The nitrogen oxides are involved through their ability to convert RO_2 (or HO_2) to RO (or OH), and also through their role as a source of ozone via nitrogen dioxide photolysis. In a vastly simplified form this can be represented by:

$$\left.\begin{matrix} NO_2 \\ \text{or} \\ RCHO \end{matrix}\right\} \xrightarrow{h\nu} OH \xrightarrow{RH} R \xrightarrow{O_2} RO_2 \xrightarrow{NO} \begin{matrix} RO \\ + \\ NO_2 \\ \downarrow \\ O_3 \text{ or } OH \end{matrix} \xrightarrow{O_2} \begin{matrix} HO_2 \\ + \\ RCHO \\ \downarrow \\ OH \end{matrix} \xrightarrow{NO} \begin{matrix} OH \\ + \\ NO_2 \\ \downarrow \\ O_3 \text{ or } OH \end{matrix} \rightarrow \text{chain continues}$$

Very complex mathematical models involving upwards of 100 reactions have been set up to simulate the reactions taking place when photochemical smog is formed and these have successfully reproduced the observed behaviour in smog-chambers where artificial urban atmospheres are irradiated under controlled conditions.

The atmospheric chemistry of sulphur-containing species has not been considered so far. In addition to sulphur compounds emitted from industrial sources, the atmosphere contains appreciable amounts of natural sulphur [215]. One estimate [216] is that only a quarter of the total sulphur generation is man-made, the remainder coming from biological sources, sea-spray and volcanoes. While natural processes take place continuously over a large part of the Earth's surface, the highly localized man-made emissions are our main concern. By far the larger proportion of these are derived from the burning of coal with smaller amounts from the combustion of oil, refinery operations and smelting of non-ferrous ores.

Sulphur compounds have a damaging effect on the respiratory tract in man: this occurs when sulphur dioxide condensed on smoke or other solid particles is inhaled and the injurious species may well be sulphate particles rather than sulphur dioxide itself. There is, however, considerable evidence that sulphur dioxide can damage plants which appear to be much more susceptible than animals. The effect of 'acid rain' on the coniferous forests, lakes and rivers of Scandinavia has been the subject of much comment and it has been claimed that the origin of the pollution has been many hundreds of kilometres from the site of the damage.

Much of the sulphur present in the atmosphere is converted to sulphates which can be detected as sulphuric acid aerosol and particles of solid salts such as ammonium sulphate. These materials are returned to the Earth's surface in rain or by dry deposition involving adsorption or chemical reaction. While they remain in the atmosphere, sulphates have a significant effect on the visibility. Other undesirable effects, for example, erosion of buildings, especially those constructed from carbonate stone, result largely from the corrosive effects of sulphuric acid.

The mechanism by which sulphur dioxide is converted to sulphate is highly complex but it appears that there are two routes:

(a) Homogeneous gas-phase oxidation, initiated photochemically.
(b) Heterogeneous oxidation, either in aqueous droplets or on aerosol particles.

and that the relative importance of each depends on the ambient conditions.

During the winter photochemical reactions are very slow and removal of sulphur dioxide is mainly by dry deposition and precipitation but in the summer, especially in polluted urban atmospheres, sulphur dioxide is oxidized by a homogeneous photochemically initiated process.

In a sunlight-irradiated atmosphere, fast free-radical reactions maintain a steady-state concentration of atoms and radicals (see above), and under these conditions the direct photolysis of SO_2 is too slow to be important. Instead sulphur dioxide reacts quickly by

$$OH + SO_2 \rightarrow HOSO_2 \tag{13.49}$$

$$HO_2 + SO_2 \rightarrow OH + SO_3 \tag{13.50}$$

$$CH_3O_2 + SO_2 \rightarrow CH_3O + SO_3 \tag{13.51}$$

The sulphur trioxide produced in Reactions 13.50 and 13.51 will hydrolyse extremely rapidly to form sulphuric acid aerosol. There is little doubt that $HOSO_2$ also ends up as the same aerosol, but the path by which this takes place is not known. In the presence of relatively high concentrations of NO_x, the aerosol also contains another compound which is probably nitrylsulphuric acid, $HOSO_2ONO_2$.

In the absence of alkenes there is no detectable dark reaction between sulphur dioxide and ozone, but when these are present, as they are in a polluted urban atmosphere, there is a fairly rapid reaction whose precise mechanism is obscure but whose products include aldehydes and sulphur trioxide [217].

The foregoing discussion has concentrated on reactions in the *troposphere*, very roughly that part of the atmosphere extending from the Earth's surface vertically upwards for about 10 km. Most of the sources of pollution mentioned so far are located close to the Earth's surface, but aircraft may generate their combustion products in the *stratosphere*, that part of the atmosphere above the troposphere and extending upwards to a height of about 50 km. The stratosphere contains a small concentration of ozone and this layer is a very important yet delicate feature of our environment. It absorbs harmful ultraviolet solar radiation in the range 280–320 nm which would otherwise reach the Earth's surface where it has been argued that it would have an adverse effect on plant and animal life. Emissions from stratospheric aircraft catalyse the conversion of the ozone in the 'shield' back to molecular oxygen. Other man-made materials which are stable in the troposphere and reach the stratosphere by diffusion include halogenated hydrocarbons and there they too undergo photochemical decomposition and may deplete the ozone layer.

The main ozone forming path is

$$O_2 + h\nu \xrightarrow{\lambda < 242\,\text{nm}} O + O \tag{13.52}$$

followed by Reaction 13.23. Reactions leading to the removal of ozone may be written

$$X + O_3 \rightarrow XO + O_2 \tag{13.53}$$

$$XO + O \rightarrow X + O_2 \tag{13.54}$$

where X = NO, H, OH or a halogen.

There have been very sophisticated attempts at modelling the ozone layer in the stratosphere but the task is difficult because of lack of knowledge of the detailed chemistry and many of the rate constants, and also because both chemistry and transport have to be taken into account. The generally accepted view is that the problem of the destruction of the ozone shield is probably not as serious as at first thought and, given the time scale and potential magnitude of the effect, steps can be taken to improve the situation while at the same time extending our knowledge of what is going on in the atmosphere [218,219].

13.2 COMBUSTION HAZARDS

Wherever large quantities of flammable materials are used or stored, there is a potential hazard from their combustion. From the early stages of the design of such installations, the potential hazards are evaluated and every possible step is taken to minimize the likely damage which would result from an accident or equipment malfunction. A number of standard tests have been devised to give the designer some indication of the danger inherent in any material. Most of these are empirical in nature and they include measurement of auto-ignition temperatures, flammability limits, flash points,* stability to heat, impact sensitivity and the like. Although many of these properties depend markedly on the way the test is carried out, when they are performed in a standardized manner they provide useful information on the relative dangers associated with each material [220,221]. The fundamental aspects of many of these topics have been discussed in earlier chapters and a proper understanding of, for example, the theory of spontaneous ignition as developed by Semenov and Frank-Kamenetskii will allow those using information from these tests to appreciate their usefulness as well as their limitations.

Hazard and operability studies now represent a considerable fraction of the total design time for any new chemical plant and the procedures, codes of practice etc. are fully discussed elsewhere [222,223]. When the plant is running, it is equally important that correct procedures are followed and, in particular, that when any repairs or alterations to the plant are contemplated, the safety implications of the work are examined most thoroughly.

Spillage of a volatile liquid such as gasoline (petrol) in the open air is followed by evaporation, and the formation of a layer of vapour whose size depends on the surface area of the liquid pool, the rate of heat transfer to the pool from its surroundings, the strength of any wind and the ambient temperature. If ignition occurs, the flame spreads through the plume and across the liquid surface at speeds of around

* The flash point is the lowest temperature at which vapour is given off from a liquid in sufficient quantity to enable ignition to take place.

5 m s^{-1}. Some of the characteristics of pool fires have been discussed in Chapter 8.

When significant quantities, say a few tonnes, of a combustible gas or liquid are released and ignited one of the following events may ensue, a *sustained fire*, a *flash fire*, a *fire ball* or a *vapour cloud explosion*. Which of these occurs depends on factors such as the vapour pressure and reactivity of the fuel, the size and manner of the release, the atmospheric conditions, the surrounding topography and the nature of the ignition source. The approximate time-scales and rates of energy release associated with these phenomena are shown in Table 13.2.

Table 13.2 The approximate time-scales and rates of energy release associated with various types of fire and explosion [224, 225].

Fire or explosion type	Characteristic time (s)	Approximate rate of energy release (MW tonne^{-1})
Sustained fire	> 100	$< 10^3$
Flash fire	10–100	10^3–10^4
Fireball	1–10	10^4–10^5
Vapour cloud explosion	< 1	$> 10^5$

Although the effects of a sustained fire are felt over a smaller distance than those of the other events listed above, its longer duration may bring about more severe localized heating. The heat transfer rates are very high (up to 200 kW m^{-2}) and very serious damage to nearby equipment and storage vessels may result.

A flash fire occurs when a release of flammable material dispersed over a wide area is ignited, the flame then spreading along the premixed vapour–air layer. The flash fire persists for only a few seconds but, near the source, it may lead to a sustained fire.

The storage of combustible gases as pressurized liquids introduces a new dimension to the discussion of potential hazards. Here, the mode of failure of the storage vessel is important. If the container splits due to, say, localized weakening, then there may be a sustained discharge of liquid or vapour or a mixture of both. Ignition of a stream of flammable material under these circumstances would tend to give a diffusion flame rather like that on a burner and although the results of the fire might be quite serious, they would be confined to the immediate vicinity of the initial failure. On the other hand, if the container is suddenly ruptured, perhaps because of the effect of an external fire, the liquid will be violently ejected in the form of vapour and small droplets. Rapid mixing with air results, and if this mixture is ignited a rapidly expanding ball of flame rises from the ground entraining more air. The thermal radiation from such a fireball is extremely intense and its effects may be felt over several hundred metres. However, because of its relatively short duration, the heating effects on massive items of plant and equipment are negligible. The main characteristics of a fireball are summarized in Table 13.3 from which it follows that the sudden release of 20 tonnes of a hydrocarbon at ground level would result in a hemispherical fireball with a diameter of about 200 m. It would last about 12 s and over this time heat would be released at a rate of 22 500 MW.

Table 13.3 Main characteristics of a fireball [226]. The mass of flammable material burning in the fireball is M kg.

Maximum diameter (sphere, m)	$5.8\ M^{1/3}$
Maximum diameter (hemisphere, m)	$7.4\ M^{1/3}$
Duration as a ground level source of radiation (s)	$0.45\ M^{1/3}$
Radiation output 20–40% of heat of combustion of the fuel mass	

Although the events described above involve ignition of the vapour-droplet cloud, it should be recognized that the rapid formation of large quantities of vapour by flash evaporation can itself produce blast waves: this phenomenon is referred to as a *boiling liquid expanding vapour explosion*, or BLEVE, and can occur with any liquid under appropriate circumstances. The subsequent ignition of a rapidly expanding cloud of flammable vapour–air mixture will lead to very rapid combustion rates and flame speeds as high as 50 m s^{-1} have been measured.

The precise circumstances which lead to a vapour cloud explosion rather than a fireball are not properly understood but when a compact, turbulent pre-mixed vapour–air cloud is partially confined by surrounding objects which promote turbulence and restrict the spread of the flame, there appears to be a greater likelihood of a vapour cloud explosion. The significance of the partial confinement is not clearly established and the phenomenon is often referred to as an *unconfined vapour cloud explosion*, or UVCE.

The explosion at Flixborough, UK, in 1974 provides a vivid illustration of what may happen if disaster strikes a large chemical

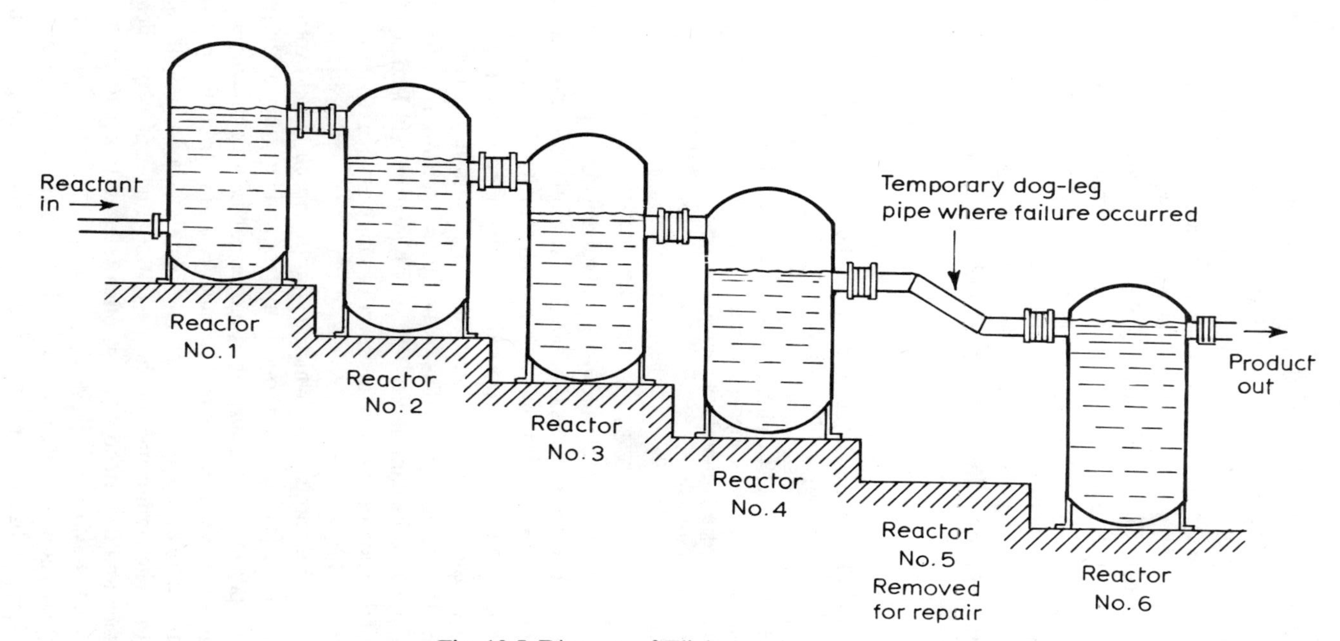

Fig. 13.5 Diagram of Flixborough reactor cascade.

plant. Here the liquid-phase oxidation of cyclohexane to cyclohexanone was carried out at 155°C and a pressure of approximately 9 atm in a cascade of six cylindrical stainless steel reactors, each of about 45 m^3 capacity and containing around 35 tonnes of cyclohexane. One of these large reactors had developed a fault and was removed, the gap between the adjacent reactors being bridged by a temporary pipe 0.5 m in diameter (Fig. 13.5). During start-up of the plant after some other minor repairs this temporary pipe ruptured releasing cyclohexane to the atmosphere at a rate exceeding 1 tonne s^{-1}. This massive escape of cyclohexane produced a cloud of spray and vapour about 200 m in diameter and, in places, 100 m high. About 45 s after the rupture, the cloud ignited and in the ensuing events 28 people died, 89 were injured and the plant was totally destroyed. Most of the houses in two villages about 1 km away were damaged, some severely, and other buildings up to 3 km distant were also affected.

The scale and extent of the damage were quite awe-inspiring and the events at Flixborough have been the subject of a most detailed enquiry and also much subsequent discussion [227]. There are many lessons to be learned regarding various aspects of the design and operation of chemical plant but in the present context the most interesting question concerns the nature of the events following ignition.

The damage from a UVCE is far in excess of that to be expected from an ordinary deflagration and this has led some to speculate that detonation may have occurred. In fact, there is no clear evidence that an unconfined vapour cloud has ever detonated, although flame speeds of 100 m s^{-1} or more are necessary to produce the overpressures required to account for the damage caused which is consistent with an exceedingly high wind rather than a shock wave [228]. The rise-time and the duration of the positive pressure from a UVCE are long compared with those of a blast wave from a high explosive generating a similar overpressure and just as with a normal blast wave, a UVCE is followed by a suction wave. The Flixborough explosion followed this pattern.

The very high flame speeds needed to explain the damage at Flixborough and other documented UVCEs imply burning velocities far in excess of those normally associated with laminar flow. Various mechanisms have been proposed to account for such high velocities including severe turbulence, absorption of radiation from hot gases, multipoint ignition and partial confinement. While ordinary meteorological conditions are unlikely to provide the necessary levels of turbulence, there are other ways in which these may be generated. In one theory [229], the flame acceleration acts differentially on pockets of heavy unburned reactant and light fully burned product leading to relative motion and the generation of turbulence within the flame. This will in turn cause further acceleration of the flame generating yet more turbulence and could produce flame speeds of the required order. Another theory [230] suggests that thermal radiation from long path lengths of hot products causes ignition of small dust particles ahead of the front and this increases the rate of flame spread through large volumes of gaseous mixtures.

One of the main difficulties in discovering the true nature of a UVCE is the absence of experimental data which can be used to test the alternative mechanisms. Almost all the data have been gathered from the analysis of damage caused by large explosions and in most cases there have been additional complications due to the presence of major structures in the vicinity causing partial confinment of the explosion, reflection of pressure waves and so on.

13.3 SUGGESTIONS FOR FURTHER READING

Bartknecht, W. (1981) *Explosions*, Springer-Verlag, Berlin.

Campbell, I.M. (1977) *Energy and the Atmosphere*, John Wiley, Chichester.

Gugan, K. (1979) *Unconfined Vapour Cloud Explosions*, Godwin, London.

Harris, R.J. (1983) *Investigation and Control of Gas Explosions in Buildings and Heating Plant*, E. & F.N. Spon, London.

Lees, F.P. (1980) *Loss Prevention in the Process Industries*, Butterworths, London.

Leighton, P.A. (1961) *The Photochemistry of Air Pollution*, Academic Press, New York.

McEwan, M.J. and Phillips, L.F. (1975) *Chemistry of the Atmosphere*, Edward Arnold, London.

Starkman, E.S. (1971) *Combustion-generated Air Pollution*, Plenum Press, New York.

Strehlow, R.A. (1976) *Prog. Energy Combust. Sci.*, **2**, 27.

References

1. Smithells, A. and Ingle, H. (1892) *Trans. Chem. Soc.*, **61**, 204.
2. Rossini, F.D. *et al.* (1952) *Selected Values of Chemical Thermodynamic Properties*, National Bureau of Standards Circular 500, Washington.
3. Wagman, D.D. *et al.* (1968 etc.) *Selected Values of Chemical Thermodynamic Properties*, National Bureau of Standards Technical Note 270, Washington.
4. Stull, D.R., Westrum, E.F. and Sinke, G.C. (1969) *The Chemical Thermodynamics of Organic Compounds*, John Wiley, New York.
5. Stull, D.R. and Prophet, H. (1971) *JANAF Thermochemical Tables*, 2nd edn., NSRDS-NBS 37, National Bureau of Standards, Washington.
6. Bett, K.E., Rowlinson, J.S. and Saville, G. (1975) *Thermodynamics for Chemical Engineers*, Athlone Press, London.
7. Hougen, O.A., Watson, K.M. and Ragatz, R.A. (1954) *Chemical Process Principles*, Part I, 2nd edn, Wiley, New York.
8. Prothero, A. (1969) *Combustion and Flame*, **13**, 399.
9. Gaydon, A.G. and Wolfhard, H.G. (1979) *Flames, Their Structure, Radiation and Temperature*, 4th edn., Chapman and Hall, London.
10. Van Zeggeren, F. and Storey, S.H. (1970). *The Computation of Chemical Equilibria*, Cambridge University Press, Cambridge.
11. Prasad, K. (1970) *Aero. J. Roy. Aeronaut. Soc.*, **74**, 757.
12. Benson, S.W. (1976) *Thermochemical Kinetics*, 2nd edn., John Wiley, New York.
13. Golden, D.M., Solly, R.K. and Benson, S.W. (1971) *J. Phys. Chem.*, **75**, 1333.
14. Emmanuel, G. (1972). *Internat. J. Chem. Kinetics*, **4**, 591.
15. Gray, B.F. and Yang, C.H. (1965) *J. Phys. Chem.*, **69**, 2747.
16. Gray, B.F. (1969) *Trans. Faraday Soc.*, **65**, 1603, 2133.
17. Gray, B.F. (1970) *Trans. Faraday Soc.*, **66**, 1118.
18. Yang, C.H. and Gray, B.F. (1969) *Trans. Faraday Soc.*, **65**, 1614.
19. Semenov, N.N. (1928) *Z. Phys.*, **48**, 571.
20. Frank-Kamenetskii, D.A. (1939) *J. Phys. Chem. Moscow*, **13**, 738.
21. Frank-Kamenetskii, D.A. (1939) *Acta Phys.-chim. URSS*, **10**, 365.
22. Frank-Kamenetskii, D.A. (1942) *Acta Phys.-chim. URSS*, **16**, 357.
23. Frank-Kamenetskii, D.A. (1945) *Acta Phys.-chim. URSS*, **20**, 729.
24. Frank-Kamenetskii, D.A. (1969) *Diffusion and Heat Transfer in Chemical Kinetics* (trans. J.P. Appleton), 2nd edn, Plenum Press, New York.

25. Tyler, B.J. and Wesley, T.A.B. (1967) *Eleventh Symposium (International) on Combustion*, The Combustion Institute, Pittsburgh, p. 1115.
26. Boddington, T., Gray, P. and Harvey, D.I. (1971) *Phil. Trans.*, **A270**, 467.
27. Boddington, T., Gray, P. and Wake, G.C. (1977) *Proc. Roy. Soc.*, **A357**, 403.
28. Boddington, T., Gray, P. and Scott, S.K. (1982). *J. Chem. Soc., Faraday Trans. II*, **77**, 801, 813.
29. Egeiban, O.M., Griffiths, J.F., Mullins, J.R. and Scott, S.K. (1982) *Nineteenth Symposium (International) on Combustion*, The Combustion Institute, Pittsburgh, p. 825.
30. Ashmore, P.G., Tyler, B.J. and Wesley, T.A.B. (1967) *Eleventh Symposium (International) on Combustion*, The Combustion Institute, Pittsburgh, p. 1133.
31. Bowes, P.C. (1973) *Combustion Institute European Symposium 1973* (ed. F.J. Weinberg), Academic Press, London, p. 142.
32. Lewis, B. and Von Elbe, G. (1961) *Combustion, Flames and Explosions of Gases*, 2nd edn, Academic Press, New York.
33. Hinshelwood, C.N. (1946) *Proc. Roy. Soc.*, **A188**, 1.
34. Hinshelwood, C.N. and Williamson, A.T. (1934) *The Reaction between Hydrogen and Oxygen*, Oxford University Press, Oxford.
35. Semenova, N. (1937) *Acta Phys.-chim. URSS*, **6**, 25.
36. Baldwin, R.R. (1956) *Trans. Faraday Soc.*, **52**, 1344.
37. Minkoff, G.J. and Tipper, C.F.H. (1962) *Chemistry of Combustion Reactions*, Butterworths, London.
38. Willbourn, A.H. and Hinshelwood, C.N. (1946) *Proc. Roy. Soc.*, **A185**, 353.
39. Davis, H.T. (1962) *Introduction to Non-Linear Differential and Integral Equations*, Dover, New York.
40. Minorsky, N. (1962) *Non-linear Oscillations*, Van Nostrand, Amsterdam.
41. Gray, B.F. (1975) Kinetics of oscillatory reactions, in *Specialist Periodical Reports, Reaction Kinetics*, Vol. 1. The Chemical Society, London, p. 309.
42. Gray, P. and Sherrington, M.E. (1977) Self-heating, chemical kinetics and spontaneously unstable systems, in *Specialist Periodical Reports, Gas Kinetics and Energy Transfer*, Vol. 2. The Chemical Society, London, p. 331.
43. Barnard, J.A. and Platts, A.G. (1972) *Combustion Sci. Technol.*, **6**, 133.
44. Gray, P. and Sherrington, M.E. (1974) *J. Chem. Soc., Faraday Trans. I*, **70**, 2338.
45. Gray, P., Griffiths, J.F., Hasko, S.M. and Lignola, P-G. (1981) *Proc. Roy. Soc.*, **A374**, 313.
46. Fine, D.H., Gray, P. and Mackinven, R. (1970) *Proc. Roy. Soc.*, **A316**, 255.
47. Coward, H.F. and Jones, G.W. (1952) *Limits of Flammability of Gases and Vapors*, US Bureau of Mines Bulletin No. 503.
48. Andrews, G.E. and Bradley, D. (1972) *Combustion and Flame,* **18**, 133.

49. Lovachev, L.A. *et al.* (1973) *Combustion and Flame*, **20**, 259.
50 Potter, A.E. (1960) *Prog. Combustion Sci. Technol.*, **1**, 145.
51. Rozlovskii, A.I. and Zakaznov, V.F. (1971) *Combustion and Flame*, **17**, 215.
52. Glassman, I. (1977) *Combustion*, Academic Press, New York.
53. Mallard, E. and Le Chatelier, H.L. (1883) *Ann. Mines*, **4**, 274, 379.
54. Tanford, C. and Pease, R.N. (1947) *J. Chem. Phys.*, **15**, 861.
55. Spalding, D.B. and Stephenson, P.L. (1971) *Proc. Roy. Soc.*, **A324**, 315.
56. Dixon-Lewis, G. (1979) *Phil. Trans.*, **A292**, 45.
57. Cherian, M.A., Rhodes, G., Simpson, R.J. and Dixon-Lewis, G. (1981) *Phil. Trans.*, **A303**, 181.
58. Tsatsaronis, G. (1978) *Combustion and Flame*, **33**, 217.
59. Spalding, D.B. (1979) *Combustion and Mass Transfer*, Pergamon, Oxford.
60. Andrews, G.E., Bradley, D. and Lwakabamba, S.G. (1975) *Combustion and Flame*, **24**, 285.
61. Damköhler, G. (1940) *Z. Elektrochemie Angewandte Phys. Chem.*, **46**, 601.
62. Abdel-Gayed, R.G. and Bradley, D. (1981) *Phil. Trans.*, **A301**, 1.
63. Bradley, D. (1982) *Oxidation Communications*, **2**, 189.
64. Longwell, J.P. and Weiss, M.A. (1955) *Industr. Eng. Chem.*, **47**, 1634.
65. Barr, J. (1954) *Fuel*, **33**, 51.
66. Jost, W. (1946) *Explosion and Combustion Processes in Gases*, McGraw-Hill, New York.
67. Burke, S.P. and Schumann, T.E.W. (1928) *Industr. Eng. Chem.*, **20**, 998.
68. Roper, F.G. (1977). *Combustion and Flame*, **29**, 219.
69. Roper, F.G., Smith, C. and Cunningham, A.C. (1977) *Combustion and Flame*, **29**, 227.
70. Roper, F.G. (1978) *Combustion and Flame*, **31**, 251.
71. Smith, S.R. and Gordon, A.S. (1956) *J. Phys. Chem.*, **60**, 759, 1059.
72. Smith, S.R. and Gordon, A.S. (1957) *J. Phys. Chem.*, **61**, 553.
73. Barnard, J.A. and Cullis, C.F. (1962). *Eighth Symposium (International) on Combustion*, Williams and Wilkins, Baltimore, p. 481.
74. Wolfhard, H.G. and Parker, W.G. (1952) *Proc. Phys. Soc.*, **A65**, 2.
75. Dows, D.A., Pimentel, G.C. and Whittle, E. (1955) *J. Chem. Phys.*, **23**, 499
76. Penney, W.G. and Pike, H.H.M. (1950) *Rept. Prog. Phys.*, **13**, 46.
77. Chapman, D.L. (1899) *Phil. Mag. (5)*, **47**, 90.
78. Jouguet, E. (1905) *J. Mathematique*, 305.
79. Jouguet, E. (1906) *J. Mathematique*, 6.
80. Jouguet, E. (1917) *Mecanique des Explosifs*, Doin, Paris.
81. Zeldovich, Y.B. (1940) *J. Exptl. Theor. Phys. USSR*, **10**, 542.
82. von Neumann, J. (1942) *OSRD Report*, No. 549.
83. Döring, W. (1943) *Ann. Physik*, **43**, 421.
84. Cher, M. and Kistiakowsky, G.B. (1958) *J. Chem. Phys.*, **29**, 506.
85. Edwards, D.H., Williams, G.T. and Breeze, J.C. (1959) *J. Fluid Mech.*, **6**, 497.
86. Edwards, D.H. (1969) *Twelfth Symposium (International) on Combustion*, The Combustion Institute, Pittsburgh, p. 819.

87. Bollinger, L.E. and Edse, R. (1961) *J. Am. Rocket Soc.,* **31**, 251, 588.
88. Shchelkin, K.I. and Troshin, Ya. K. (1965) *Gasdynamics of Combustion,* Mono Book Corp., Baltimore.
89. Strehlow, R.A. (1968) *Combustion and Flame*, **12**, 90.
90. Oran, E.S., Young, T.R., Boris, J.P., Picone, J.M. and Edwards, D.H. (1982) *Nineteenth Symposium (International) on Combustion*, The Combustion Institute, Pittsburgh, p. 573.
91. Baulch, D.L., Drysdale, D.D., Horne, D.G. and Lloyd, A.C. (1972) *Evaluated Rate Data for High Temperature Reactions*, Vol. 1, Butterworths, London.
92. Trotman-Dickenson, A.F. and Ratajczak, E. (1969) *Supplementary Tables of Bimolecular Gas Reactions*, UWIST, Cardiff.
93. Walker, R.W. (1975) A critical survey of rate constants for gas-phase hydrocarbon oxidation, in *Specialist Periodical Report, Reaction Kinetics,* Vol. 1, The Chemical Society, London, p. 161.
94. Walker, R.W. (1977) Rate constants for reactions in gas-phase hydrocarbon oxidation, in *Specialist Periodical Report, Gas Kinetics and Energy Transfer*, Vol. 2, The Chemical Society, London, p. 296.
95. Baldwin, R.R. and Walker, R.W. (1979) *J. Chem. Soc. Faraday Trans. 1,* **75**, 140.
96. Baulch, D.L. and Campbell, I.M. (1981) Gas Phase Reactions of Hydroxyl Radicals, *Specialist Periodical Report, Gas Kinetics and Energy Transfer*, Vol. 4, The Royal Society of Chemistry, London, p. 137.
97. Kaufman, M. and Sherwell, J. (1983) *Prog. Reaction Kinetics,*
98. Baldwin, R.R. and Walker, P.W. (1972) Branched-chain reactions: the hydrogen–oxygen reaction, in *Essays in Chemistry* (eds J.N. Bradley, R.D. Gillard and R.F. Hudson), Vol. 3, Academic Press, London.
99. Dixon-Lewis, G. and Williams, D.J. (1977) The oxidation of hydrogen and carbon monoxide, in *Comprehensive Chemical Kinetics* (eds C.H. Bamford and C.F.H. Tipper), Vol. 17, Elsevier, Amsterdam.
100. Bradley, J.N. (1967) *Trans. Faraday Soc.*, **63**, 2945.
101. Kurzius, S.C. and Boudart, M. (1968) *Combustion and Flame*, **12**, 477.
102. Baldwin, R.R., Rossiter, B.N. and Walker, R.W. (1969) *Trans. Faraday Soc.*, **65**, 1044.
103. Foo, K.K. and Yang, C.H. (1971) *Combustion and Flame*, **17**, 223.
104. Barnard, J.A. and Platts, A.G. (1972) *Combustion Sci. Technol.*, **6**, 177.
105. Jachimowski, C.J. and Houghton, W.M. (1971) *Combustion and Flame*, **17**, 25.
106. Cheaney, D.E., Davies, D.A., Davis, A., Hoare, D.E., Protheroe, J. and Walsh, A.D. (1959) *Seventh Symposium (International) on Combustion,* Butterworths, London, p. 183.
107. Hoare, D.E. and Walsh, A.D. (1954) *Trans. Faraday Soc.,* **50**, 37.
108. Hoare, D.E., Ting-Man Li. and Walsh, A.D. (1967) *Eleventh Symposium (International) on Combustion*, The Combustion Institute, Pittsburgh, p. 879.
109. Antonik, S. and Lucquin, M. (1968) *Bull. Soc. Chem. Fr.*, **10**, 4043.
110. Barnard, J.A. and Harwood, B.A. (1974) *Combustion and Flame*, **22**, 35.
111. Semenov, N.N. (1935) *Chemical Kinetics and Chain Reactions*, Oxford University Press, Oxford.

112. Knox, J.H. (1959) *Seventh Symposium (International) on Combustion*, Butterworths, London, p. 122.
113. Knox, J.H. (1959) *Trans. Faraday Soc.*, **55**, 1362.
114. Knox, J.H. (1960) *Trans. Faraday Soc.*, **56**, 1225.
115. Semenov, N.N. (1944) *CR Acad. Sci. URSS*, **43**, 342.
116. Baldwin, R.R., Bennett, J.P. and Walker, R.W. (1980) *J. Chem. Soc. Faraday Trans. I*, **76**, 2396.
117. Baldwin, R.R. and Walker, R.W. (1981) *Eighteenth Symposium (International) on Combustion*, The Combustion Institute, Pittsburgh, p. 819.
118. Baldwin, R.R., Hisham, M.W.M. and Walker, R.W. (1982) *J. Chem. Soc. Faraday Trans. I*, **78**, 1615.
119. Baldwin, R.R. and Walker, R.W. (1973) *Combustion and Flame*, **21**, 55.
120. Baldwin, R.R., Bennett, J.P. and Walker, R.W. (1980) *J. Chem. Soc. Faraday Trans. I*, **76**, 1075.
121. Ashmore, P.G. (1963) *Catalysis and Inhibition of Chemical Reactions*, Butterworths, London, p. 344.
122. Norrish, R.G.W. and Foord, S.G. (1936) *Proc. Roy. Soc.*, **A157**, 503.
123. Barat, P., Cullis, C.F. and Pollard, R.T. (1972) *Proc. Roy. Soc.*, **A329**, 433.
124. Bradley, J.N. and Durden, D.A. (1972) *Combustion and Flame*, **19**, 452.
125. Warnatz, J. (1981) *Eighteenth Symposium (International) on Combustion*, The Combustion Institute, Pittsburgh, p. 369.
126. Dixon-Lewis, G. (1981) *First Specialists Meeting (International) of the Combustion Institute*, The Combustion Institute (French Section), Bordeaux, p. 284.
127. Boers, A.L., Alkemade, C.T.J. and Smit, J.A. (1956) *Physica*, **22**, 358.
128. Polanyi, M. (1932) *Atomic Reactions*, Williams & Northgate, London.
129. Laidler, K.J. (1955) *The Chemical Kinetics of Excited States*, Oxford University Press, Oxford.
130. Shuler, K.E. (1953) *J. Phys. Chem.*, **57**, 396.
131. Gaydon, A.G. and Wolfhard, H.G. (1951) *Proc. Roy. Soc.*, **A208**, 63.
132. Calcote, H.F. and King, I.R. (1955) *Fifth Symposium (International) on Combustion*, Reinhold, New York, p. 423.
133. Payne, K.G. and Weinberg, F.J. (1959) *Proc. Roy. Soc.*, **A250**, 316.
134. Palmer, H.B. and Cullis, C.F. (1965) *Chemistry and Physics of Carbon* (ed. P.L. Walker), Dekker, New York.
135. Homann, K.H. (1967) *Combustion and Flame*, **11**, 265.
136. Cullis, C.F. (1975) *Amer. Chem. Soc. Symp.*, 21, *Petroleum-derived Carbons*, p. 348.
137. Bittner, J.D. and Howard, J.B. (1981) *Eighteenth Symposium (International) on Combustion*, The Combustion Institute, Pittsburgh, p. 1105.
138. Calcote, H.F. (1981) *Combustion and Flame*, **42**, 215.
139. Sheinson, R.S. and Williams, F.W. (1973) *Combustion and Flame*, **21**, 221.
140. Barnard, J.A. and Watts, A. (1969) *Twelfth Symposium (International) on Combustion*, The Combustion Institute, Pittsburgh, p. 365.
141. Barnard, J.A. and Watts, A. (1972) *Combustion Sci. Technol.*, **6**, 125.

142. Barnard, J.A. and Brench, A.W. (1977) *Combustion Sci. Technol.*, **15**, 243.
143. Halstead, M.P., Prothero, A. and Quinn, C.P. (1971) *Proc. Roy. Soc.*, **A322**, 377.
144. Halstead, M.P., Kirsch, L.J., Prothero, A. and Quinn, C.P. (1975) *Proc. Roy. Soc.*, **A346**, 515.
145. Yang, C.H. and Gray, B.F. (1969) *J. Phys. Chem.*, **73**, 3395.
146. Yang, C.H. (1969) *J. Phys. Chem.*, **73**, 3407.
147. Frank-Kamenetskii, D.A. (1940) *Zh. Fiz. Khim*, **47**, 30.
148. Salnikov, I.E. (1949) *Zh. Fiz. Khim*, **23**, 258.
149. Gray, P., Griffiths, J.F., Hasko, S.M. and Lignola, P-G. (1981) *Proc. Roy. Soc.*, **A374**, 313.
150. Gray, P., Griffiths, J.F., Hasko, S.M. and Lignola, P-G. (1981) *Combustion and Flame*, **43**, 175.
151. Yang, C.H. (1974) *Combustion and Flame*, **23**, 97.
152. Jacobs, P.W.M. and Whitehead, H.M. (1969) *Chem. Rev.*, **69**, 551.
153. Jacobs, P.W.M. and Pearson, G.S. (1969) *Combustion and Flame*, **13**, 419.
154. Jacobs, P.W.M. and Gilbert, R. (1971) *Can. J. Chem.*, **49**, 2827.
155. Eyring, H., Powell, R.E., Duffey, G.H. and Parkin, R.B. (1949) *Chem. Rev.*, **45**, 69.
156. Cook, M.A. (1958) *The Science of High Explosives*, Reinhold, New York.
157. Bawn, C.E.H. (1948) *Ministry of Supply Report* A.C. 10068/LFC58.
158. Huggett, C. (1956) Combustion of solid propellants, in *Combustion Processes* (eds B. Lewis, R.N. Pease, and H.S. Taylor), Oxford University Press, Oxford, p. 514.
159. Fristrom, R.M. and Westenberg, A.A. (1965) *Flame Structure*, McGraw-Hill, New York.
160. Long, V.D. (1964) *J. Inst. Fuel*, **37**, 522.
161. Williams, A. (1973) *Combustion and Flame*, **21**, 1.
162. Williams, A. (1976) *Combustion of Sprays of Liquid Fuels*, Elek, London.
163. Herzberg, M. (1973) *Combustion and Flame*, **21**, 195.
164. De Ris, J. and Orloff, L. (1972) *Combustion and Flame*, **18**, 381.
165. Orloff, L. and De Ris, J. (1982) *Nineteenth Symposium (International) on Combustion*, The Combustion Institute, Pittsburgh, p. 895.
166. Burgoyne, J.H. Roberts, A.F. and Quinton, P.G. (1968) *Proc. Roy. Soc.*, **A308**, 39.
167. Burgoyne, J.H. and Roberts, A.F. (1968) *Proc. Roy. Soc.*, **A308**, 55, 69.
168. Roberts, A.F. and Quince, B.W. (1973) *Combustion and Flame*, **20**, 245.
169. Smith, I.W. (1982) *Nineteenth Symposium (International) on Combustion*, The Combustion Institute, Pittsburgh, p. 1045.
170. Mulcahy, M.F.R. (1978) The combustion of carbon in oxygen, in *The Metallurgical and Gaseous Fuel Industries*, The Chemical Society, London, p. 175.
171. Fordham, S. (1980) *High Explosives and Propellants*, 2nd edn, Pergamon, Oxford.
172. Livingston, C.W. (1956) *Sixth Annual Drilling and Blasting Symposium*, University of Minnesota, p. 44.
173. Livingston, C.W. (1956) *Colo. Sch. Min. Quart.*, **51**, 1.

174. Munroe, C.E. (1885) *Van Nostrand's English Magazine*, **32**, 1.
175. Munroe, C.E. (1888) *Am. J. Sci.*, **36**, 48.
176. Neumann, M. (1910/1911) *Westfalisch Anhaltische Springstaffe, AG*, German patent 249630/1910, British patent 28030/1911.
177. Bray, K.N.C. (1959) *J. Fluid Mech.*, **6**, 1.
178. *Handbook of Chemistry and Physics*, 39th edn, Chemical Rubber Publishing Co., Cleveland, Ohio, 1957.
179. Siegel, B. and Schieler, B. (1964) *Energetics of Propellant Chemistry*, John Wiley, New York.
180. Haskell, W.W. and Bame, J.L. (1966) *SAE Trans.*, **74**, 772.
181. American Petroleum Institute (1956) *Research Project 45*, 18th Annual Report.
182. Walsh, A.D. (1963) *Ninth Symposium (International) on Combustion*, Academic Press, New York, p. 1046.
183. Affleck, W.S. and Fish, A. (1967) *Eleventh Symposium (International) on Combustion*, The Combustion Institute, Pittsburgh, p. 1003.
184. Hirst, S.L. and Kirsch, L.J. (1980) in *Combustion Modeling in Reciprocating Engines* (ed. J.N. Mattavi and C.A. Amann), Plenum Publishing Corporation, New York.
185. Ellison, R.J., Harrow, G.A. and Hayward, B.M. (1968) *J. Inst. Petrol.*, **54**, 243.
186. Norrish, R.G.W. (1959) *Seventh Symposium (International) on Combustion*, Butterworths, London, p. 203.
187. Erhard, K.H.L. and Norrish, R.G.W. (1960) *Proc. Roy. Soc.*, **A259**, 297.
188. Norrish, R.G.W (1965) *Tenth Symposium (International) on Combustion*, The Combustion Institute, Pittsburgh, p. 11.
189. Downs, D., Griffiths, S.T. and Wheeler, R.W. (1961) *J. Inst. Petrol.*, **47**, 1.
190. Zimpel, C.F. and Graiff, L.B. (1967) *Eleventh Symposium (International) on Combustion*, The Combustion Institute, Pittsburgh, p. 1015.
191. Rao, V.K. and Prasad, C.R. (1972) *Combustion and Flame*, **18**, 167.
192. Homer, J.B. and Hurle, I.R. (1972) *Proc. Roy. Soc.*, **A327**, 61.
193. Dryer, F.L. (1977) *Sixteenth Symposium (International) on Combustion*, The Combustion Institute, Pittsburgh, p. 279.
194. Elias, A.R. (1978) *Proc. International Symposium on Automotive Technology and Automation*, Vol. 2, Wolfsburg, p. 215.
195. Greeves, G., Khan, I.M. and Onion, G. (1977) *Sixteenth Symposium (International) on Combustion*, The Combustion Institute, Pittsburgh, p. 321.
196. Heywood, J.B. (1976) *Prog. Energy Combust. Sci.* **1**, 135.
197. Ham, R.W. and Smith, H.M. (1951) *Ind. Eng. Chem.*, **43**, 2788.
198. De Gray, A. (1922) *Rev. Metall.*, **19**, 645.
199. Yates, J.G. (1983) *Fundamentals of Fluidized-bed Chemical Processes*, Butterworths, London.
200. Davidson, J.F. and Harrison, D. (1963) *Fluidised Particles*, Cambridge University Press, Cambridge.
201. Thring, M.W. (1952) *Fuel*, **31**, 355.
202. Vulis, L.A. (1961) *Thermal Regimes of Combustion* (trans. M.D. Friedman), McGraw-Hill, New York.

203. Houghton, J.T. (1979) *Phil. Trans.* **A290**, 515.
204. Starkman, E.S. (1969) *Twelfth Symposium (International) on Combustion*, The Combustion Institute, Pittsburgh, p. 598.
205. Baulch, D.L. and Drysdale, D.D. (1974) *Combustion and Flame*, **23**, 215.
206. Hayhurst, A.N. and Vince, I.M. (1980) *Prog. Energy Combustion Sci.*, **6**, 35.
207. Hayhurst, A.N. and Vince, I.M. (1983) *Combustion and Flame*, **50**, 41.
208. *Ford Energy Report* (1981) **2**, 5.
209. Adamczyk, A.A., Kaiser, E.W., Lavoie, G.A. and Isack, A.J. (1983) *Combustion and Flame*, **52**, 1.
210. Henein, N.A. (1976) *Prog. Energy Combustion Sci.*, **1**, 165.
211. Chamberlain, A.C., Heard, M.J., Little, P. and Wiffen, R.D. (1979) *Phil. Trans.*, **A290**, 577.
212. Smith, F.B. and Hunt, R.D. (1979) *Phil. Trans.*, **A290**, 523.
213. Logan, J.A., Prather, M.J., Wofsy, S.C. and McElroy, M.B. (1978) *Phil. Trans.*, **A290**, 187.
214. Kerr, J.A., Calvert, J.G. and Demerjian, K. (1972) *Chem. Br.*, **8**, 252.
215. Cullis, C.F. (1978) *Chem. Br.*, **14**, 384.
216. Garland, J.A. (1977) *Proc. Roy. Soc.*, **A354**, 245.
217. Cox, R.A. (1979) *Phil. Trans.*, **A290**, 543.
218. Thrush, B.A. (1979) *Phil. Trans.*, **A290**, 505.
219. Poppoff, I.G., Whitten, R.C., Turco, R.P., and Capone, L.A. (1978) *NASA Reference Publication 1026*, NASA.
220. Bretherick, L. (1979) *Handbook of Reactive Chemical Hazards*, 2nd edn, Butterworths, London.
221. Sax, N.I. (1979) *Dangerous Properties of Industrial Materials*, 5th edn, Van Nostrand Reinhold, New York.
222. Lees, F.P. (1980) *Loss Prevention in the Process Industries*, Vol. 1 and 2, Butterworths, London.
223. *A Guide to Hazard and Operability Studies*, Chemical Industries Association, London (1977).
224. McQuaid, J. and Roberts, A.F. (1982) *Institution of Chemical Engineers 50th Jubilee Symposium.*
225. Roberts, A.F. (1982) *Inst. Chem. Eng. Symp. Ser.*, **71**, 181.
226. Roberts, A.F. (1981/82) *Fire Safety J.*, **4**, 197.
227. *Report of the Court of Enquiry on the Flixborough Disaster*, HMSO, London (1975).
228. Roberts, A.F. and Pritchard, D.K. (1982) *J. Occupational Accidents*, **3**, 231.
229. Bray, K.N.C. and Moss, J.B. (1981) *First Specialists Meeting (International) of the Combustion Institute*, The Combustion Institute (French Section), Bordeaux, p. 7.
230. Moore, S.R. and Weinberg, F.J. (1983) *Proc. Roy. Soc.*, **A385**, 373.

Answers to problems with numerical solutions

Chapter 1

1.

Fuel	MJ kg^{-1} Reactant	MJ kg^{-1} Fuel	MJ m^{-3}
H_2	3.86	143	12.8
CH_4	2.93	55.6	39.7
C_2H_6	2.92	52.0	69.6
C_3H_8	2.91	50.5	99.1
C_2H_2	3.37	50.0	58.0
C_2H_4	3.07	50.4	63.0
$CH_3OH(g)$	3.08	23.9	34.1
$CH_3OH(l)$	2.92	22.6	1.79×10^4

(The maximum thermal energy is released when water is produced in the *liquid* state.)

2.

	Estimates (J mol^{-1}K^{-1}) C_V	Estimates (J mol^{-1}K^{-1}) C_p	C_p from Table 1.2 (J mol^{-1}K^{-1}) 298 K	1500 K	Mean
CO_2	37.4	45.7	37.14	58.36	47.8
H_2O	37.4	45.7	33.63	46.69	40.2
N_2	25.0	33.0	29.09	34.52	31.8

3.

	Constant volume	Constant pressure
O_2	8115 K	6687 K
Air	2984 K	2371 K

4. 8×10^{-17}, 8×10^{-6}, 2×10^{-3}, 5×10^{-2}

5. 13.2, 718, 381, 1016, 0.40 mol dm^{-3} s^{-1}
6. $10^{18.6}$ dm$^{3/2}$ mol$^{-1/2}$ s^{-1}; 358 kJ mol^{-1}
   C_2H_4 and H_2; CH_4 and C_4H_{10}.
   $[CH_3] = 9 \times 10^{-13}$ mol dm^{-3}; $[H] = 1 \times 10^{12}$ mol dm^{-3};
   $[C_2H_5] = 4 \times 10^{-10}$ mol dm^{-3}; 7.5×10^{-7} mol dm^{-3} s^{-1};
   6×10^{-11} mol dm^{-3} s^{-1}

7. $D \propto (\text{pressure})^{-1}$, $(\text{temperature})^{3/2}$, $(\text{molar mass})^{-1/2}$
   $\lambda \propto (\text{pressure})^{0}$, $(\text{temperature})^{1/2}$, $(\text{molar mass})^{1/2}$

8. $466\,\text{m}\,\text{s}^{-1}$; $1.03 \times 10^{-7}\,\text{m}$; $1.60 \times 10^{-5}\,\text{m}\,\text{s}^{-2}$; $1.48 \times 10^{-2}\,\text{W}\,\text{m}^{-1}\,\text{K}^{-1}$.
   $Le = 0.714$ and $1/\gamma$

Chapter 2

1. 1520 m, 1.92 m
2. 10.9 K
3. $\delta_c = 3.32$, $\Delta T_c = 19.4\,\text{K}$
5. $0.5\,\text{s}^{-1}$

Chapter 3

1.

	d_Q(mm)	d_T(mm)
H_2	0.84	1.29
CH_4	2.45	3.77
C_3H_8	2.41	3.72
C_2H_2	0.66	1.02

2. 24.1 mm
3. $1300\,\text{mol}\,\text{m}^{-3}\,\text{s}^{-1}$
   1 mm
4. Increased by a factor of 4.4
6. Changed by a factor of (a) 0.7 (b) 1.4

Chapter 4

H_2 3878 K, $2505\,\text{m}\,\text{s}^{-1}$, 22.3 atm
CH_4 2730 K, $1825\,\text{m}\,\text{s}^{-1}$, 15.8 atm

Chapter 5

2. $1.54 \times 10^{8}\,\text{m}^3\,\text{mol}^{-1}\,\text{s}^{-1}$; $67.3\,\text{kJ}\,\text{mol}^{-1}$
3. $67.5\,\text{kJ}\,\text{mol}^{-1}$
4. 0.40
5. 0.42

Chapter 6

4.

		300° C	500° C	800° C
(a)	1,4p	4.4	1.1×10^4	6.3×10^6
	1,4t	750	2.9×10^5	3.5×10^7
	1,5p	129	7.4×10^4	1.2×10^7
(b)	1,7p	1.6×10^3	1.3×10^5	4.7×10^6
	1,6s	1.3×10^4	9.8×10^5	3.2×10^7
	1,5s	1.5×10^3	3.5×10^5	2.7×10^7
	1,4s	64	5.7×10^5	1.3×10^7
	1,4p	1.45	3.8×10^3	2.1×10^6
	1,4s	64	5.7×10^4	1.3×10^7
	1,5s	1.5×10^3	3.5×10^5	2.7×10^7
	1,6p	437	8.9×10^4	6.4×10^6
	1,4s	129	1.1×10^5	2.6×10^7
	1,5p	129	7.4×10^4	1.2×10^7

Chapter 7

4.2×10^{10}, 7.1×10^8, 2.9×10^{13}, 3.7×10^{16} ions m^{-3}

Chapter 8

1. 1.10 $mm^2 s^{-1}$; 0.68 s, 0.91 s
2. (a) 7.64, (b) 33.4
3. 2.10 ms, 1.5 mm

Index